FINDING A PLACE FOR ENERGY

SITING COAL CONVERSION FACILITIES

Frank J. Calzonetti

Department of Geology and Geography
and
Regional Research Institute
West Virginia University
Morgantown

with

Mark S. Eckert

Grand Junction
Colorado

RESOURCE PUBLICATIONS
IN GEOGRAPHY

Copyright 1981

by The

Association of American Geographers
1710 Sixteenth Street N.W.
Washington, D.C. 20009

Library of Congress Card Number 81-69236
ISBN 0-89291-147-6

Library of Congress Cataloguing in Publication Data

Calzonetti, Frank J.
  Finding a place for energy.

  (Resource publications in geography)
  Bibliography: p.
  1. Energy facilities — Location. 2. Coal-fired power plants — Location. 3. Coal gasification. 4. Coal liquefaction. I. Eckert, Mark S., 1948-
II. Title. III. Series.
TJ163.2.C34   338.6'042   81-69236
ISBN 0-89291-147-6 (pbk.)   AACR2

Publication Supported by the A.A.G.

Graphic Design by D. Sue Jones and CGK

Printed by Commercial Printing Inc.
State College, Pennsylvania

Cover Photograph: Courtesy of Monongahela Power Company

Foreword

One decade ago, hardly anyone in the industrial nations anticipated the events about to unfold in the world energy economy. For some nations, fuel shortages and energy price increases of the 1970s brought social disruption and debilitating balance-of-payments burdens. For several, new energy discoveries occurred at an opportune time. As the world's largest energy consumer, the United States was nearly unique in experiencing major difficulties as the world energy situation vacillated from shortage to surfeit, while having significant underdeveloped and unexploited domestic energy resources. Anticipation that we could reach a mythical potential of energy self-sufficiency in the 1980s has turned to disillusionment. We now realize that domestic fuels can only abate, not eliminate our international energy dependency. For example, salvation through the vast U.S. coal reserves — in the traditional eastern production areas and in newly-opened western basins — has yet to occur. It is clear, however, that these resources will be important in our ultimate transition to a non-fossil energy economy, whether based on atomic or renewable energy systems.

This book examines one major element in the use of domestic coal resources in ameliorating our dependency on imported fuel, the siting of coal conversion facilities. Coal *can* be consumed directly, at least for heating. But it is a bulky, dirty fuel requiring more effort than most people accustomed to automated central heating systems are willing to expend. Moreover, coal used directly provides little potential for lighting or mobile power. Thus coal conversion is a virtual necessity. Only by consuming coal as electricity or synthetic liquid or gaseous fuels can we maintain the kind of end-use energy consuming system we have structured during this century. Indeed, moderate to large coal conversion facilities can utilize advanced technologies to diminish some of the harmful effects of coal combustion, including atmospheric pollution, management of solid wastes, and disposal of reject heat.

Before we can realize our potential for use of coal, we must have the conversion facilities. Power plants require sites. The process of identifying, licensing, and developing energy facility sites is the focus of this volume. Here we learn about the siting process in practice and in law, as well as the contribution of geographical research toward structuring tomorrow's energy systems. The most important kind of facility addressed is the electrical generating plant, but siting implications of coal gasification and liquefaction are also explored. Whether one agrees that the next generation of coal conversion facilities may be the *last* of their genre, as suggested by authors Frank Calzonetti and Mark Eckert, this volume argues persuasively for a geographic contribution to siting. However, does a geographical role in locating a power plant mean using the traditional tools of industrial location and marketing analysis? Or do legal requirements and political realities suggest a new locational approach?

This book is the second of the **Resource Publications in Geography** series, sponsored by the Association of American Geographers, a professional organization whose purpose is to advance studies in geography and to encourage the application of geographic research in education, government, and business. Through its contribution to our economic life, geographic research affects all of us.

Resource Publications in Geography traces its origins to the Association's Commission on College Geography whose *Resource Papers* were launched in 1968. Eventually 28 papers were published under sponsorship of the Commission through 1974 with assistance from the National Science Foundation. Continued NSF support after completion of the Commission's work permitted the *Resource Papers for College Geography* to meet the original series goals over an additional four years and sixteen volumes:

> The Resource Papers have been developed as expository documents for the use of both the student and the instructor. They are experimental in that they are designed to supplement existing texts and to fill a gap between significant research in American geography and readily accessible materials. The papers are concerned with important concepts or topics in modern geography and focus on one of three general themes: geographic theory; policy implications; or contemporary social relevance. They are designed to implement a variety of undergraduate college geography courses at the introductory and advanced level.

The popularity and usefulness of the two series suggested the importance of their continuation after 1978 once a self-supporting basis for their publication had been established.

For this second volume of **Resource Publications** in 1981, the original goals remain paramount. However, they have been broadened to include the continuing education of professional geographers as well as communication with the public on contemporary issues of geographic relevance. This monograph was developed, printed and distributed under the auspices of the Association, whose members served in advisory and review roles during its preparation. The ideas presented, however, are those of the authors and do not imply AAG endorsement.

For our students, we hope that this book will stimulate thought about the energy system that sustains us and about policy issues in energy facility siting. Reference to Earl Cook's *Energy: The Ultimate Resource* (a 1978 *Resource Paper*) may provide helpful background. For our friends in electric ultilities and energy firms, we trust that this volume will provide perspectives useful in your planning. For members of legislative and regulatory bodies, as well as citizen interest groups, we trust that this book will help to unravel some of the complexity of energy facility siting. For fellow geographers, we intend that this sharing of disciplinary perspectives and insight will enhance our professional or local role in siting decisions.

C. Gregory Knight
The Pennsylvania State University
Editor, Resource Publications in Geography

Resource Publications Advisory Board

George W. Carey, *Rutgers University*
James S. Gardner, *University of Waterloo*
Charles M. Good, Jr., *Virginia Polytechnic Institute and State University*
Mark S. Monmonier, *Syracuse University*
Risa I. Palm, *University of Colorado*
Thomas J. Wilbanks, *Oak Ridge National Laboratories*

Preface and Acknowledgements

The United States, as indeed most nations, faces pressing energy problems. Few nations are capable of satisfying projected energy demand without expensive energy importation. Two major approaches exist to remedy this situation. Demand can be curtailed to reduce the gap between domestic supply and energy needs. Alternatively, domestic energy supplies can be increased. In most countries, it is recognized that the severity of the problem mandates both approaches.

Although there may be a national consensus that new energy conversion facilities are needed, little consensus exists as to where these new energy facilities should be sited. Problems and delays are encountered in siting almost every new energy facility. There are many reasons why it is difficult to site and to construct new energy facilities. An uncertain economic and regulatory climate, energy demand growth rates below previous expectations, construction material delivery problems, expanding environmental protection programs, increased citizen and interest group activism, and uncertain large scale technologies are some of the principal factors. Indeed, the problems encountered in siting energy facilities may be *the* limiting factor preventing the U.S. from increasing its production of apparently plentiful domestic energy sources.

In siting energy facilities, locational choices must be made on the basis of different and often competing criteria. In some respects, energy facility siting is a special case of industrial location geography, using traditional tools and expertise in that domain. However, an energy facility siting literature is evolving quite independently of location theory. Practitioners are often not able to transfer simplified theoretical contributions to problems as complex as locating major energy facilities.

This book surveys problems that occur in siting facilities that convert coal into more useful forms. These are coal-fired power plants, coal liquefaction plants, and coal gasification plants. We emphasize these facilities because coal is our most important source of electricity, and may become significant in supplying gaseous and liquid fuels as well. Proposed coal conversion facilities require large resource inputs (coal, water, land), produce undesirable residuals (gaseous, liquid, and solid waste), and are often viewed as noxious. However, the nature and intensity of a facility's impacts depend upon its site.

Professional geographers have been increasingly interested in energy problems. Not only are many colleges and universities offering geography courses addressing energy topics, but many instructors teaching economic, industrial, resource, and political geography are looking at energy problems as a fruitful topic of interest. Geographers are increasingly evident in the energy industry. This book is intended for all students of energy facility siting. We hope that it will enhance geographers' contributions toward an important contemporary problem.

Our interest in this topic can be traced to 1976 when we were members of an energy facility siting research team at the University of Oklahoma. This research team was led by Dr. Thomas J. Wilbanks who was then Chairman of the Department of Geography at Oklahoma. The research conducted at Oklahoma formed the basis of the energy facility siting section of the Science and Public Policy Program's *Energy From the West* study which is heavily used in this manuscript. We have maintained contact with Dr. Wilbanks in writing this book. Notorious for his relentless critiques, Dr. Wilbanks provided the authors with a thorough review of an earlier draft of this manuscript. His time is greatly appreciated.

The editor of the AAG *Resource Publication* series, C. Gregory Knight, was of great help at every stage of development of this book. His experience in energy geography was helpful in focusing the scope of the book and his detailed critiques of the individual chapters strengthened and tightened the manuscript. Michael R. Greenberg of Rutgers University also provided helpful comments on the final draft.

Many other people were of great help in the preparation of this book. David Hawley and Mary Traeger drafted the illustrations. Jo Ann Calzonetti, a librarian at West Virginia University, provided professional reference support. Lucinda Robinson, the Assistant Director of the Regional Research Institute, West Virginia University, provided much support and technical guidance in the preparation of the manuscript. Most importantly, however, the contribution of the typists at the Regional Research Institute and Department of Geology and Geography, West Virginia University, is gratefully acknowledged. These fine people are Jean Gallaher, who typed the initial draft, and Debbie Benson, Kathy Minyon, and Carla Uphold who typed revisions and the final copy.

Frank J. Calzonetti
Mark S. Eckert

Contents

1	**ENERGY FACILITY SITING**	1
	Coal Conversion Facility Siting	3
	Problems in Facility Siting	4
	Patterns of Energy Facility Location	7
2	**IMPACTS OF COAL CONVERSION**	8
	Components of Coal Conversion Facilities	8
	Coal-Fired Power Plants	8
	Coal Liquefaction Plants	10
	Coal Gasification Plants	12
	Comparative Impacts of Coal Conversion	13
	"Best-Fit" Location Planning	15
	Coal Resource Transportation Systems	17
	Siting and the Distribution of Impacts	17
3	**PERMITS AND APPROVALS**	20
	Facility Siting Problems	20
	The Federal Role	22
	The State Role	26
	Municipal and County Oversight	28
	Licensing Delays	28
	Lack of Coordination	30
4	**CENTRALIZED OR DECENTRALIZED ENERGY?**	31
	Decentralized Energy	32
	Centralized Energy	34
5	**THE DECISION ENVIRONMENT IN ENERGY FACILITY SITING**	39
	Location Theory and the Behavior of Large Organizations	39
	Organizations Siting Coal Conversion Facilities	40
	The Decision Environment	41
	Uniform Delivered Price	41
	Capital Cost and Risk Minimization	42
	Least-Cost Siting Approaches	43

6	**ANALYZING SITING OPTIONS**	47
	Site Screening Methods	47
	Spatial Allocation Models	50
	Resolving Locational Conflict	53
	Increased Citizen Involvement	54
	Mitigation and Compensation	55
7	**GUIDING ENERGY SUPPLY THROUGH THE END OF THE FOSSIL-FUEL ERA**	58
	Adjusting Patterns of Development	58
	Regional Shifts in Energy Supply	61
	Conclusion	62
	BIBLIOGRAPHY	63

List of Figures

1	Location of the Kaiparowits Power Project	2
2	The Energy Facility Siting Process	5
3	Major Components of a Coal-Fired Power Plant	9
4	Centralized and Decentralized Energy Alternatives	33
5	Location of Announced Coal Conversion Facilities	34
6	Median Capacity of Coal-Fired Power Plant Units	35
7	National Electric Reliability Council Regions	38
8	The Least-Cost Siting Approach	44
9	Standard and Compensatory Screening Approaches	49
10	Maryland Power Plant Site Screening	51

List of Tables

1	Energy Facility Siting Tasks	6
2	Outputs from a Proposed Coal Liquefaction Facility	12
3	Residuals of Coal Conversion	14
4	Water Consumption by Conversion Technology	15
5	Technological and Locational Factors Affecting Energy Development Impacts	16
6	Important Siting Criteria for Utilities	21
7	Major Permits Required for a Coal Gasification Facility in North Dakota	23
8	Major Permits Required for a Coal Liquefaction Facility in West Virginia	24
9	Alternative Actions to a Coal Liquefaction Facility in West Virginia	25
10	State Power Plant Siting Legislation	26
11	North Dakota Exclusion and Avoidance Areas	27
12	Causes of Power Plant Delay	29
13	U.S. Coal-Fired Electrical Generating Capacity	36
14	Electric Utility Generation in the U.S.	40
15	Desired Attributes for Energy Facility Location	48
16	Outcomes of Approaches to Conflict Situations	53
17	State Energy Severance Tax Collections, 1979	56
18	Typical Attributes of Small and Large Power Plants	59

1

Energy Facility Siting

In 1961, a group of investor-owned utilities led by Southern California Edison began planning a 5,000 megawatt power plant on the Kaiparowits Plateau of southern Utah. The developers envisioned that this $500 million project, the largest in the country, would burn low sulfur coal to generate electricity for Arizona, Nevada, and California (Myhra 1977:25). After a decade of conflict, permit filing, and project revisions, the developers withdrew their application in 1976. Although the size of the project was reduced from 5,000 to 3,000 megawatts, the price tag of the project escalated from $500 million to $3.7 billion (Myhra 1977:25). According to the utilities, the project failed because of rising costs and increasing uncertainty resulting from years of delay, opposition, and "red tape." The executive vice president of Southern California Edison remarked that the project was "beaten to death by the environmental interests" (OECD 1980:54).

The Kaiparowits project was a part of an ambitious plan conceived by a consortium of 23 utilities to produce 36,000 megawatts of electricity from the Four Corners area (OECD 1980:49). The 2,075-Mw Four Corners plant which began operations in 1963 was the first of many facilities to be sited in the region. Utility planners had the support of the governor of Utah and many citizens of Utah who viewed 35,000 new jobs, an additional payroll of $100 million, and yearly tax revenues of $28 million as a path to a prosperous future (Myhra 1977:25). Support also came from utility organizations and even the Federal Energy Office.

The pristine nature of the region was one of the reasons the Kaiparowits project encountered such difficulties. While the Kaiparowits site itself was not viewed as being particularly sacred, the project would be centered among the nation's largest concentration of national parks (Figure 1). Grand Canyon, Bryce, Zion, Arches, and Capital Reef would have been within 250 miles of the facility as would other scenic areas such as Monument Valley and the Canyonlands, all known for their pristine vistas.

The Sierra Club, Audubon Society, Friends of the Earth, Wilderness Society, Environmental Defense Fund, and other groups worked against the project. The predecessor Navajo Plant, one of the largest point sources of pollution in the nation, alerted these groups to the potential of serious impacts of new facilities. These groups were able to focus adverse national publicity on the Kaiparowits proposal. They also initiated court action which delayed the licensing procedure. When water rights needed for cooling were granted by Secretary of the Interior Hickel in 1969, it appeared that the facility would soon be sited. However, the passage of the National Environmental Policy Act later in the year caused further delay. The Sierra Club quickly sued, forcing the

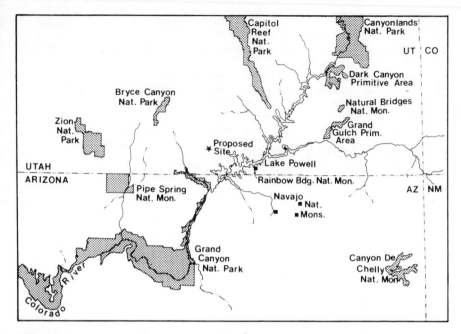

FIGURE 1   LOCATION OF THE KAIPAROWITS POWER PROJECT

Department of the Interior to submit an environmental impact statement (EIS), a requirement of the new law. Even after receiving approval from the Department of the Interior, the project still needed 220 permits and authorizations from 42 federal, state, and local agencies (Myhra 1977:25). The outcome of these delays was an escalation in the price of the project, which was estimated in 1975 to be rising at $1 million per day.

After the impact statement was prepared, the Sierra Club then filed a petition with the California Public Utilities Commission to determine whether the facility was essential to maintain reliable electric service. With increased scrutiny on the part of the California Public Utilities Commission in assessing the need for the facility, and, in addition, other delays and public opposition, the utilities withdrew their application. Although the utilities had already invested over $20 million into preliminary analysis for the project, they withdrew rather than face continuing uncertainty.

Siting problems, blamed by the utilities for dooming Kaiparowits, may have been overshadowed by more fundamental economic factors. Rock (1977) points out that when the project was conceived the utilities involved were experiencing rapidly growing electricity demand. By the mid-70s, when the controversy was reaching its peak, demand growth had slowed considerably. Kaiparowits was planned to meet electricity sales increases of close to 10 percent per year, as experienced during the 60s and early 70s. Because of slumps in electricity sales after 1973, the prime developer, Southern California Edison, reduced its forecast of sales for 1984 by 30 percent (Rock 1977:250). At this time, the City of Los Angeles experienced the most dramatic electricity demand *decreases* in the nation's history. In 1973, when it was realized that contracted Arab oil would not be delivered, the city had initiated an energy conservation plan to prevent electricity brown-outs. By 1974, this plan resulted in a 12 percent drop in sales

(Stobaugh and Yergin 1979:144-145). As demand growth levels softened, the price of the project continued to rise. Rock (1977:250) summarizes what he felt was the principal reason for the withdrawal of the Kaiparowits power plant application:

> The demise of Kaiparowits was jointly determined by the reduction in demand for electricity and the relatively high cost of Kaiparowits power. . . . Social costs and benefits, however, did not enter the Kaiparowits' decision-matrix because the energy consortium (or its financiers) was the only decision-maker. Profit was the main criterion.

The Kaiparowits situation raises some important questions concerning the significance of siting problems in altering utility plans. It is not known how important the role of delays, conflicts, poor public image, and red tape was in causing the demise of the project. Power company officials may have exploited these developments as a convenient excuse to scale down the original proposal and withdraw from a project that was becoming increasingly uneconomic in the face of declining electricity sales growth rates. In doing so, the power companies may have focused attention on siting problems that they believed should be eliminated while avoiding an evaluation of their own judgments in power planning. Their position, moreover, was widely reported in the popular press (*Business Week* 1976). The case has been used by many state and federal officials as evidence that efforts should be made to streamline the siting process and to limit the length of environmental impact statements and associated public hearings. Conversely, Kaiparowits has not been regarded as evidence that utility planning and capacity building programs must be reformed.

Coal Conversion Facility Siting

The siting of energy facilities is a pressing problem likely to grow even more acute in the future. This book focuses on the problems involved in siting coal conversion facilities (power plants, coal liquefaction plants, and coal gasification plants) likely to be called upon to provide a substantial part of the nation's energy needs. Furthermore, the locational question may be the paramount issue in deploying coal conversion technologies. Nuclear power is opposed regardless of location. Wherever a nuclear plant is proposed, conflicts occur — many national interest groups have strong antinuclear platforms and can mobilize local support to fight a proposed facility. Coal facilities, while constrained by aspatial economic considerations, are not so vigorously attacked. The location of the coal facility determines its competition for resources, environmental and health impacts, and socioeconomic disruption. These locationally varying factors confront energy planners and are the key to public acceptance of a coal facility. Some major coal facilities have been sited and built without significant problems. Other identical facilities proposed in different areas have aroused national uproar and prolonged licensing delays, as was the case with Kaiparowits.

Using coal as a source of electricity and gaseous and liquid fuels raises coal conversion facility siting as an issue worthy of significant attention. Many energy studies, irrespective of their long-term solutions to energy problems, maintain that coal must make some contribution in the coming decades to their alleviation. Coal is viewed as the one resource whose production could be increased quickly enough to offset declining domestic oil and gas reserves. The major use for coal will be in the generation of electricity. In 1980, over 80 percent of the nation's coal was consumed by electric utilities. The momentum to use coal as a boiler fuel has increased in relation to the uncertainties of nuclear power which until recently was viewed as the principal fuel for future electrical generation. A study by the U.S. Department of Energy (1980: vii 18) estimated a loss of approximately 10 percent of the nation's potential 1985 electricity production because of nuclear plants not being built as anticipated. In addition, a study by Komanoff reported in *Science* (Norman 1981) indicates the escalation in the capital costs of nuclear power plants will further increase the attractiveness of coal systems.

A future coal supply system would require a large number of power plants, coal liquefaction plants, and coal gasification plants sited throughout the nation. Planning the location of these facilities requires careful study. They require prodigious quantities of local resources (some of which are scarce, such as water in the western United States). They also generate a wide range of gaseous, liquid, and solid wastes which may result in undesirable impacts to the local area or contribute to region-wide or national pollution problems (acid rainfall). In some cases, groups and individuals object to the aesthetic impact of a facility (stacks and cooling towers may be an insult to an otherwise natural vista). In addition, servicing the facility may strain transportation systems. Furthermore, many object because construction workers may cause serious socioeconomic disruption to local communities. On the other hand, a new energy facility may bring rare employment and tax revenue opportunities to a small town.

Siting these facilities will continue to raise serious problems. These include problems that arise in the siting of particular facilities as well as the long-term patterns of energy facility location that arise as a consequence of many individual siting decisions. While particular siting problems capture the attention of popular literature, the consequences of many single decisions may have profound effects on the pattern of the nation's energy supply. Geographers have added their expertise and skills both in addressing individual siting problems and in evaluating and planning the nation's emerging energy supply system.

Problems in Facility Siting

The process of siting a new energy facility is illustrated in Figure 2. The four decision steps in siting a facility at a particular location are:

(1) Determining the need for a new energy facility;
(2) Choosing the appropriate energy technology to satisfy that need;
(3) Choosing a location for the particular facility; and
(4) Gaining final approval for the facility.

If the facility is not approved, then decisions must be made to reevaluate the need for the plant, the technology, or the site. If the facility is approved and constructed, then it contributes to the nation's energy supply.

Many problems involved in siting new facilities are shown in Table 1, where a wide range of decision-makers and academic disciplines are involved. Determining the need for new energy facilities involves forecasting energy demand. This activity is done at different scales and for different purposes. The federal government forecasts national and regional energy demand. Energy utilities use their own forecasts as the basis for determining when new facilities should be built. Some state governments (public utility commissions) also are involved in forecasting studies used to verify those compiled by energy utilities and energy corporations.

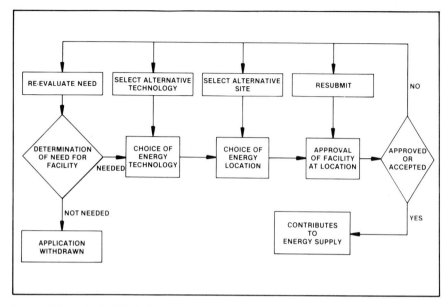

FIGURE 2   THE ENERGY FACILITY SITING PROCESS

Once the need for a new facility has been established, the next task is to select an energy technology. The size and type of facility is matched with the anticipated need. These decisions are generally made by the utilities or energy corporations in conjunction with energy technology vendors.

Choosing the site for the energy facility involves the expertise of a wider range of individuals and requires working with more decision-makers. The first stage of selecting a project site is to confine the search to a region of interest, followed by a more careful evaluation of potential sites. The manner in which sites are selected depends upon siting and environmental laws in the state under consideration. Some states have prescribed potential locations whereas others evaluate sites after they have been selected by the energy developer. Geographers have been involved in both aspects of the site determination problem and have even helped states to devise energy facility siting plans. Detailed analysis by civil engineers and geologists is necessary before a utility or energy corporation will make a commitment to build a facility at a particular location.

After the site has been selected, it is necessary to gain approval for the facility. This requires presenting information at public hearings and securing a long list of federal, state, and local approvals and authorizations. At this point conflicts generally arise between those who advocate development and those opposing the facility. Opposition may be restricted to strong statements at public hearings by concerned individuals or can involve court proceedings, demonstrations, or acts of violence. In some cases, the undesirable consequences of the proposed facility raised by opposing interests are serious enough to cause modifications to the facility, relocation, reapplication for an operating license, or withdrawal of the application.

The siting process is seldom as organized or sequential as this discussion may suggest. Part of the problem in siting energy facilities stems from the fact that the decision-making process is more confused than orderly and is inconsistent in sequence

TABLE 1   ENERGY FACILITY SITING TASKS

TASK	DECISION-MAKING	ACADEMIC REALMS
Determining the Need for New Energy Facilities — Forecasting Energy Demand	Energy Utility and Energy Corporation System Planners Federal Energy Agencies State Energy Agencies and Public Utility Commissions	Economics Operations Research Management Science
Choosing an Energy Technology — Matching Energy Technology, Size, and Type to Forecasted Demand	Energy Utility and Energy Corporation System Planners and Engineers Federal Energy Agencies State Energy Agencies and Public Utility Commissions Energy Technology Vendors	Economics Operations Research Power Engineering Mechanical Engineering
Choosing an Energy Facility Location — Determining a Region of Interest	Energy Utility and Energy Corporation System Planners, Engineers, and Locational Analysts Federal Energy Agencies State Energy Agencies and Public Utility Commissions	Economics Operations Research Transportation Research Geography Power Engineering
Determining an Energy Facility Site	Energy Utility and Energy Corporation System Planners, Engineers, Locational Analysts, Geologists, and Environmental Specialists Federal Energy and Environmental Agencies State Energy Agencies and Public Utility Commissions Regional and Local Planning Agencies Citizen Interest Groups	Mechanical, Power, and Civil Engineering Geology Geography (Locational Analysis and Environmental Impacts) Environmental Sciences Law Political Science
Gaining Approval of a Facility at a Location — Securing Necessary Licenses and Approvals; Dealing with Locational Conflict	Energy Utility and Energy Corporation Legal Department, Environmental Impact Specialists, and Public Information Personnel Federal Environmental and Health Agencies State Energy, Environmental, and Health Agencies State Public Utility Commissions Regional and Local Planning Agencies Citizen Interest Groups Environmental Groups	Law Political Science Public Relations Environmental Sciences

or content. For example, it is generally after a site for a particular energy facility has been announced that questions are raised about the need for the facility. Those opposing a particular facility may suggest energy conservation or solar energy as an alternative to a new coal facility.

The nature of siting problems depends upon the perspective of the observer. Citizen groups may feel that decisions are made hastily without sufficient input from interests other than the developer. Power company officials find other reasons to fault the process. Siting problems routinely identified by power company executives are:

(1) The need for a new energy facility is disputed;
(2) The technology or design of the facility is questioned by members of the public and regulatory agencies;
(3) The site chosen for the facility is viewed as inappropriate by many groups and individuals;
(4) The procedure of filing applications for required permits and authorizations is too lengthy, overlapping, and cumbersome; and
(5) The project is often subject to assault by opposing interests who are able to foster further delay and inconvenience to the energy developer by extending the permitting process and instigating legal battles.

The prototype situation can be characterized as one in which a decision to build a facility at a particular location is challenged along a wide front of separate issues, as was the proposed Kaiparowits power plant.

Patterns of Energy Facility Location

Another important aspect of siting involves determining those factors that influence locational decisions and evaluating the overall impact of the resulting patterns of energy facilities. Location decisions, besides being affected by traditional industrial location factors (raw materials, market, labor), are heavily influenced by federal environmental legislation, state siting laws, and utility and energy corporation policy. Some federal laws, such as the Clean Air Act, have been important in altering facility location patterns in the United States. Unless these siting implications of environmental legislation are understood, the resulting unanticipated pattern of energy facilities may be undesirable. States also can encourage or discourage developers from siting within their boundaries. A corporation may wish to avoid becoming involved in a state that has strict licensing procedures but may be attracted to a state that has a less restrictive program. Local levels of government can also promote or discourage energy developers by providing land and services for new facilities or establishing restrictive ordinances.

These factors may result in energy facilities becoming concentrated in certain areas of the country. It is important to evaluate the resultant pattern of energy facilities that can be so strongly influenced by the current situation, because they will be producing energy well into the 21st century.

An important part of understanding siting problems is an awareness of the nature of large coal conversion facilities. The next chapter discusses these technologies and the reasons why they are considered undesirable neighbors. We will then turn our attention to siting problems, potential patterns of development, and strategies in addressing contemporary energy location problems.

2

Impacts of Coal Conversion

Coal conversion facilities may affect a local area in undesirable ways. Licensing procedures for approving these facilities are geared to safeguarding health and the environment by specifying construction, design, and location standards. However, coal conversion facilities cannot be made completely clean or safe. Disputes may arise over the potential undesirable impacts of the facility compared to its potential benefits. Many groups support local energy development because of increased energy supply, more jobs, and tax revenues; others doubt that these positive developments will offset the damage posed by such energy development. Part of the communication problem in facility siting disputes results from a lack of understanding of coal conversion technologies, although disputes do occur between equally informed experts. In this chapter we provide brief sketches of a coal-fired power plant, coal liquefaction plant, and coal gasification plant. We then compare the impacts that may arise from the deployment of these technologies. Finally, we show that adjusting location can noticeably alter the nature and severity of a particular facility's impacts, demonstrating the importance of siting in reducing the consequences of coal-based energy systems.

Components of Coal Conversion Facilities

Coal-Fired Power Plants

Coal-fired power plants provide approximately 40 percent of the nation's electricity (U.S. Department of Energy, Energy Information Administration 1981). A wide variety of power plant designs exist, but essentially they consist of a boiler, a steam turbine, a generator, and cooling systems. In addition, modern power plants employ air pollution control equipment. We will outline the operation of a typical power plant that uses wet scrubbers and electrostatic precipitators for air pollution control. A 1,000 MWe power plant consumes about 2.5 million tons of coal annually.* Efficiencies of these plants are limited by energy conversion losses and pollution abatement equipment. The typical power plant equipped with scrubbers converts about 38 percent of the coal's chemical energy into electricity with the remainder released into the environment as reject heat

---

*MWe = megawatts (electric). Mwh = megawatt - hours (1000 kilowatt - hours, Kwh). A 1000 MWe plant at full operating capacity for one hour produces 1000 Mwh. One gigawatt = 1000 Mw.

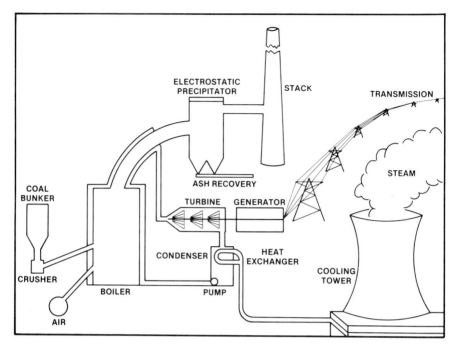

FIGURE 3   MAJOR COMPONENTS OF A COAL-FIRED POWER PLANT

(National Research Council 1979: 161). The operation of the power plant is best understood in terms of its three energy conversion stages, cooling system, and stack gas cleaning operations. Figure 3 is a simplified sketch of a modern coal-fired power plant.

The first energy conversion stage involves the conversion of the chemical energy in the coal into heat energy through coal combustion. Pulverized coal is blown into the furnace, setting up a cyclonic burning pattern. The heat generated is transferred to water which is circulating in pipes around the furnace. When heated, the water is converted to high pressure steam. Most boilers convert from 80 to 90 percent of the heat from coal combustion into steam. The second energy conversion stage occurs when this high pressure steam is directed to spin the blades of a turbine, converting steam energy into mechanical energy. The thermodynamic efficiency of the turbine is approximately 50 percent. This sets up the final energy conversion stage, mechanical into electrical energy. This is accomplished by a generator which is attached to the turbine. The motion of an electrical conductor through a magnetic field produces the alternating current that can be transmitted into an electrical power grid (Radian Corporation 1976:239; Science and Public Policy Program 1975:12-6 to 12-11; Stoker, Seager and Capener 1975:161). Thus, the potential overall efficiency of a modern coal-fired unit is the product of the coal conversion (.85), turbine (.50), and generation efficiencies (.98), or approximately 40 percent without scrubbers (Sears, Zemansky, and Young 1980:350). Only a few power plants are able to operate at even 40 percent efficiency, which means that the remainder of the coal input is released either as pollutants or as waste heat. As emphasized by Commoner (1977), most centralized energy systems have poor thermodynamic efficiencies, contributing to the pollution problems of this energy path. The billowing white clouds one sees from power plant cooling towers is low temperature waste heat being released into the environment. This heat is part of the coal's unused chemical energy that will remain unavailable to perform further work. Some countries — Sweden, Poland, and the Soviet Union — use this power plant waste heat for district heating, providing inexpensive steam heat for nearby communities.

The cover shows a facility on the Ohio River with wet cooling towers, although other cooling systems are commonly used (dry cooling towers, cooling ponds, once-through cooling). Cooling towers were designated as the "best available control technology" for thermal pollution control in the 1972 Amendments to the Federal Water Pollution Control Act and are now used on over 80 percent of new power plants (Reynolds 1980:368). Cooling water is in a separate system from the water that circulates through the boiler. After the high pressure steam transfers energy to the turbine, it is directed into a condenser. In the condenser, heat is transferred from the steam to the separate cooling water system by means of a heat exchanger. The cooling action of the heat exchanger and lower pressures in the condenser convert the steam back into water, which is then pumped back into the boiler to repeat the heat transfer cycle. The cooling system water from the heat exchanger is pumped to the cooling tower where it dissipates heat into the air by evaporation.

Electrostatic precipitators are used to remove particulate air pollutants from the stack gases. The May 1979 Clean Air Act regulations now require scrubbers on all new power plants to remove gaseous sulfur dioxide, irrespective of the quality of the coal burned. These regulations are currently being reviewed; anthracite has already been exempted from these provisions. The stack gases first enter the precipitator which removes between 91 and 99 percent of the particulates (Radian Corporation 1976: 257). The precipitator operates by imposing a high electric field on wires and tubes on which ionized particulates collect. Periodically these wires and tubes are vibrated, releasing particulates which fall in the form of fly ash. Fly ash comprises approximately 80 percent of the total ash produced by such a power plant. The remaining 20 percent is bottom ash and slag. Whereas bottom ash and slag can be sold as a road treatment or landfill material, only about 15 percent of the nation's fly ash is sold. If projected coal facilities come on-stream and oil burning units switch to coal, over 157 million tons of utility ash will be generated annually in the United States, making ash the fourth largest solid material produced in the nation, exceeded only by stone, sand and gravel, and coal (Faber 1976). Most operators remove ash from the power plant site in trucks for land disposal.

In wet scrubbing systems, stack gases come into contact with a lime solution. Sulfur dioxide reacts with the solution and is removed. Stack gases are released, and sulfur-rich sludge is either disposed of or recovered for other uses (Stoker, Seager, and Capener 1975:171), such as a raw material in the production of phosphorus fertilizers. A typical wet scrubbing system on a facility burning two percent sulfur coal produces 200 pounds of dry sludge for each ton of coal burned (National Research Council 1979:165).

Coal Liquefaction Plants

There are no commercial coal liquefaction facilities in the United States, although several pilot plants have been in operation; several commercial-scale demonstration plants are in the planning stages. Federal support for coal liquefaction was boosted with the signing of the $20 billion synthetic fuel bill by President Carter in 1980 which formed the U.S. Synthetic Fuels Corporation (Farney 1980). Recent action by the Reagan Administration has diminished the outlook for the development of this technology. Coal liquefaction is currently being used in South Africa at a plant owned by South African Oil, Coal, and Gas Corporation of Sasolburg. Nazi Germany developed a coal liquefaction capability. Although a significant contribution to the wartime economy, it was modest compared to the size of modern liquefaction facilities, such as the proposed installation near Morgantown, West Virginia. This facility, a demonstration plant, uses one of many coal liquefaction processes. Although it may not be the model for a U.S. coal liquefaction capability, this facility was the furthest in its planning stages of all coal liquefaction projects in the nation. The final environmental impact statement was submitted and construction contracts were made. However, a decision was made in June 1981 to stop work on this project. Nevertheless, it does demonstrate siting problems that may arise in the development of coal liquefaction technology.

The Morgantown facility was to be a solvent-refined coal (SRC-II) demonstration plant which was to convert 6,000 tons of coal per day into liquid fuels and other by-products. If the facility appeared to be commercially feasible after a two-year trial period, the operation would have been expanded to accommodate 30,000 tons of coal per day and produce 100,000 barrels of liquid fuel (U.S. Department of Energy 1980:xxii). The cost of this project, estimated at $1.4 billion, was to be borne by the U.S. government, the Pittsburgh and Midway Coal Mining Company (a subsidiary of Gulf Oil Corporation), and the governments of Japan and West Germany. The Germans and the Japanese pledged to finance 50 percent of the cost of the facility, but withdrew from the project because of a lack of U.S. government support. Gulf Oil's contribution would have been approximately $15 million (although it invested an additional $85 million into the development of the solvent-refined coal process); the U.S. federal government was to be responsible for all other expenses (Pasztor 1981). If proven feasible, Gulf could have purchased the facility. The federal government's goal in this project was to reduce the technical and economical risks associated with new commercial activities so that energy technologies could develop much faster than would ordinarily occur.

Solvent-refined coal is one of four coal liquefaction approaches. The other three are indirect liquefaction, pyrolysis, and catalytic liquefaction. The facility in operation in Sasolburg, South Africa, uses the indirect liquefaction approach which is believed to be most promising by many (National Research Council 1979:178-179). The commercial viability of the indirect approach is even recognized in the final environmental impact statement for the SRC-II plant (U.S. Department of Energy 1981).

In the proposed Morgantown facility, high-sulfur eastern coal was to be converted into a low sulfur coal within a hydrogen atmosphere (Table 2). As the coal dissolves, it picks up hydrogen. The coal is converted into liquid and gaseous products. This liquid solution is then drained and filtered to remove ash and undissolved coal (Science and Public Policy Program 1975:97). Vapors produced in the process are separated and treated. The methane-rich vapor is upgraded to produce pipeline quality synthetic gas. The overall efficiency of the facility, measured in terms of the heating value of the products compared to the feedstock coal, is 65 percent (U.S. Department of Energy 1980b:2-9). However, if this fuel is used to power electrical generating facilities, the total efficiencies in terms of energy delivered to the customers is quite low. The overall efficiency of using liquefied coal for electric heat would be approximately 24 percent compared to an overall efficiency of 41 percent if the resulting liquid fuel was used for oil heat.

TABLE 2 OUTPUTS FROM A PROPOSED COAL LIQUEFACTION FACILITY (per day)

OUTPUT	QUANTITY
Products	
Substitute natural gas	41.8×10^6 ft^3
Mixed liquid butane product	1461 bbl
Liquid propane product	2070 bbl
Light fuel oil (nominal boiling range 350-600°F, 14.4° API)	4900 bbl
Heavy fuel oil (nominal boiling range 600-950°F, 7.2° API)	5386 bbl
By-products	
Anhydrous ammonia	30 short ton
Elemental sulfur	135 short ton
Tar acids	43 bbl
Plant fuel products	
Naptha (nominal boiling range 150-350°F, 31 API)	2364 bbl
Synthetic fuel gas	30.8×10^6 ft^3

Source: U.S. Department of Energy 1981.

Coal Gasification Plants

Coal gasification facilities also convert coal into a more useful form of energy. The process can produce synthetic natural gas (SNG) that is equivalent to natural gas derived from underground reservoirs. Synthetic natural gas contains almost no sulfur, carbon monoxide, or free oxygen. Once produced, synthetic gas can be added to the present natural gas pipeline system and transported throughout the nation to serve particular areas.

The idea of producing gas from coal is not new. Many cities were lighted by low Btu "town gas" from the 19th century to World War II (National Research Council 1979:173).* This gas, a mixture of nitrogen, carbon monoxide, and hydrogen, was produced by passing air and steam through beds of hot coal (Stoker, Seager, and Capener 1975:182). Inexpensive natural gas from domestic reservoirs, transported by the pipelines constructed during and after World War II, quickly displaced synthetic gas.

Gas can be derived from coal by means of *in-situ* operations in which gas is drawn from fractured and heated coal seams or by surface operations. *In-situ* gasification is not a proven technology, nor are its impacts certain. However, there are numerus technologies for chemically synthesizing gas from coal in surface facilities. All processes involve the addition of hydrogen to heated coal or the removal of carbon from coal, because the hydrogen content of coal is on the order of 5 percent compared to 25 percent found in natural gas and intermediate- or high-Btu gas.

A high-Btu gasification project in North Dakota, principally sponsored by American Natural Resources Company of Detroit, is closest to commercial operation of all U.S. coal gasification projects. The facility uses the Lurgi high-Btu process which requires three principal ingredients: coal, hydrogen, and oxygen. Local lignite, available in vast

---

*Btu = British thermal unit, a unit of energy.

quantities, is crushed in a vessel where it is heated in an oxygen-rich atmosphere. Gas produced here is upgraded in a hydrogenation step in which hydrogen is added in the form of steam from Missouri River water. The resulting gas is cleaned of carbon dioxide and hydrogen sulfide impurities. To produce high-Btu gas, the product is passed over a catalyst (nickel compounds are likely) to upgrade the gas to pipeline-quality methane (Science and Public Policy Program 1975:72). A commercial-scale facility would be composed of a number of gasifier reactors capable of producing about 10 million cubic feet of synthetic gas per day. A large facility could cost an amount equal to the total assets of a gas utility but provide only 10 percent of its supplies.

Comparative Impacts of Coal Conversion

In evaluating the impacts of an energy facility, one must recognize the manner in which a given technology interacts with a particular location. Each energy technology produces residuals which are defined as:

> . . . by-products that an activity, process or technological alternative produces in addition to its primary product. Residuals include particulates, gases, solid and liquid wastes, accidents and death, and land consumption, all or some of which might produce significant environmental impacts where they occur (Science and Public Policy Program 1975:14-1).

The impacts resulting from the employment of an energy technology depend upon the interaction of residuals and local environmental conditions. A power plant may release large quantities of sulfur dioxide, carbon monoxide, and particulates into the atmosphere at a particular location. The impact may be frequent "poison fogs" if low-level atmospheric inversions are common in the area, or the area may have few air pollution episodes if tall stacks and windy conditions disperse pollutants. Knowledge of the way in which a facility's residuals are likely to interact with an area provides a basis for planning the location of energy facilities to minimize undesirable impacts.

Table 3 compares the air, water, and solid waste residuals for coal-fired power plants, coal gasification plants, and coal liquefaction plants processing different coals. There is a great range of residuals between technologies as well as a considerable variation within a particular technology depending upon the coal used. Sulfur dioxide emissions from a power plant burning eastern coal may be over three times those of the same facility using western coal. Such regional variations in the quality of coal explain why power plant operators in the East purchase midwestern or western coal to mix with local coal so that their emissions are within federal standards.

Coal-fired electric power plants emit more gaseous residuals than coal gasification or coal liquefaction. These pollutants include particulates, sulfur oxides (SO_x), nitrogen oxides (NO_x), hydrocarbons, and carbon monoxide (CO). Thus it is more difficult to site power plants to meet ambient air pollution regulations in areas where only a small increase in air degradation is permitted. As noted by White et al. (1977:31), synthetic fuel facilities can usually meet all federal and state standards (except for hydrocarbons in the case of coal liquefaction), depending upon the location of the facility and the effectiveness of pollution abatement equipment. Some air pollution control regions cannot accept even modest increases in additional point sources of pollution, and air pollution control may be a significant factor constraining siting.

TABLE 3 RESIDUALS OF COAL CONVERSION (TONS/10^{12} Btu)

POLLUTANTS	Coal-Fired Power Plants[a]		Solvent Refined Coal		Lurgi High-Btu Gasification	
	Eastern Coal[b]	Western Coal[c]	Northern Appalachian Coal	Northwest Coal	Central Coal	Northwest Coal
Water						
Dissolved Solids	0	0	0	52.4	43.1	0
Suspended Solids	12.5	12.5	0.017	0	0.9	0
Organics	5.5	5.5	0.003	0	0.426	0
AIR						
Particulates	50.0	35.0	3.25	3.49	3.65	2.05
NO_x	300.0	390.0	88.2	88.5	73.3	76.9
SO_x	250.0	80.0	14.3	4.81	36.8	5.59
Hydrocarbons	6.5	8.0	0.295	0.296	1.22	1.28
CO	21.0	27.0	2.5	2.51	4.07	4.27
Aldehydes	NC	NC	0.3	0.27	0.448	0.292
SOLIDS (10^3 tons)	149	76	47.1	34.6	52.7	37.3

NC = not considered
[a] Assumes facilities use wet limestone scrubbing.
[b] Eastern coal is assumed to be 3.0 percent sulfur and 14.4 percent ash.
[c] Western coal is assumed to be 0.8 percent sulfur and 8.4 percent ash.
Source: Science and Public Policy Program (1975).

The ash and sulfur content of the coal is critical in determining the nature of liquid effluents and solid wastes. Residuals from coal gasification are almost exclusively ash, whereas effluents from electric power plants are comprised of almost equal amounts of ash and sludge from flue gas desulfurization (White et al. 1977:49). The proposed SRC-II facility would generate over 250,000 tons of solid wastes per year at the demonstration stage (U.S. Department of Energy 1981:2-9) and would produce over one million tons per year if it became fully commercial. A 3,000-MWe power plant complex using Northern Appalachian coal will also generate over a million tons of solid wastes per year (Calzonetti and Elmes 1981).

The labor intensity of a particular facility also is a contributing factor to the air pollution impacts associated with energy development. The *Energy from the West* study (White et al. 1977) found that peak ground-level concentrations of particulates, NO_2, and hydrocarbons produced by energy-related urban development were higher in most cases than those produced by the energy facilities themselves.

The water intensiveness of a facility also is an important characteristic which is often crucial in determining its acceptance in an area. Table 4 summarizes consumptive water requirements for coal conversion facilities. Electric power plants are the most water-intensive facilities, whereas Lurgi gasification is the least water intensive. Approximately 80 percent of the total water requirements for these facilities is for cooling.

Using wet/dry instead of wet cooling could reduce water consumption by 72 percent (White et al. 1977:49). As noted in the 1979 White study (1979:99), facility location can be the critical factor determining water consumption of energy development:

> . . . water requirements for a Lurgi facility in the Four Corners area can be about twice that required for the same facility in the northern Great Plains. This is because of the low moisture content of the coal in New Mexico, the fact that the Lurgi process accepts wet coal, the high ash New Mexico coal requiring more water for disposal, and the need for supplemental irrigation to reclaim the land.

Although water problems are most acute in the western states, developers are finding that water availability is also becoming an important public issue in eastern states as well.

The overall water needs of energy facilities are modest compared to other water uses on a national scale. Irrigation accounts for 76 percent of the total water consumed in the 48 contiguous states, compared to less than two percent for fossil and nuclear energy supply systems (National Research Council 1979:197).

"Best-Fit" Location Planning

Knowledge of technological and locational factors can be a useful tool in planning the location of new energy facilities (White et al. 1979). Table 5 outlines the significant technological and locational factors that should be taken into account when planning the location of energy facilities. Labor intensity of the facility affects all four of the impacts discussed: air quality, water availability and quality, socioeconomic, and ecological. A larger population results in higher automobile emissions, a need for more water and an increase in sewage treatment capability, a wider variety of culture and lifestyles (which create problems where small homogeneous communities are affected), and greater land-use infringement on the surrounding wildlife habitat. Among the coal conversion technologies, on a unit energy output basis, electric power generation produces the most air impacts and uses the largest quantities of land and water. Coal gasification and liquefaction require larger work forces and result in the greatest population-related impacts.

TABLE 4   WATER CONSUMPTION BY CONVERSION TECHNOLOGY

TECHNOLOGY	WATER CONSUMED	
	Gallons/10^6 Product Btu	Acre-Feet Per Year[a]
Coal Fired Power Plants	127-159	23,880-29,820
Lurgi Gasification	14-24	3,310- 5,640
Synthoil Liquefaction	15-19	9,230-11,750

[a] For a 3,000 MWe power plant at 70 percent load factor; for 250 million cubic feet per day gasification facilities at 90 percent load factor; 100,000 barrels per day coal liquefaction facilities at 90 percent load factor. Exact consumption varies according to the location of the facility and the coal used.
Source: White et al. 1977:51.

TABLE 5 TECHNOLOGICAL AND LOCATIONAL FACTORS AFFECTING ENERGY DEVELOPMENT IMPACTS[a]

IMPACTS	FACTORS	
	Technological	Locational
AIR QUALITY	Emission quantities	Energy resource characteristics
	Labor intensiveness	Meterological conditions
		Topography
		Class of PSD area
WATER AVAILABILITY & QUALITY	Water requirements	Water availability
	Labor intensiveness	Water quality
	Amount and composition	Energy resource characteristics
	Type of cooling	Aquifer characteristics
		Capacity of existing wastewater treatment facilities
SOCIOECONOMIC	Labor intensity	Community size and location
	Capital intensity	Capabilities of existing institutions
	Scheduling of construction	Historical outmigration
		Local labor force characteristics
		Local financial conditions
		Culture and lifestyles of the area
ECOLOGICAL	Land requirements	Climate
	Water requirements	Topography
	Labor intensity	Soils
	Air emissions	Plant and animal communities

PSD = Prevention of Significant Deterioration
[a]Several sets of factors would be involved in large-scale developments that include more than one technology, such as a coal mine and a power plant at the same site.
Source: White *et al.* 1979

Locational characteristics which infuence the level of impact include: local topography, air pollution dispersion potential, background levels of air pollutants, meteorological conditions, proximity of the site to pristine areas such as national parks, community size and location, available work force, characteristics of the local economy, characteristics of the resources, water availability and quality, and plant and animal communities.

The size of the host community is crucial in determining the degree of socioeconomic impact. Small towns generally have limited planning capabilities and inadequate public facilities and services to accommodate the needs of a large number of incoming workers. Siting energy facilities near larger towns may be preferable.

The air and water impacts resulting from energy facilities also vary with location. Air impacts depend critically on local meteorology, topography, existing air quality, and the nearness of Class I Prevention of Significant Deterioration (PSD) areas. Water impacts differ with the amount and quality of available water, including both surface and ground-water. Air and water impacts can be reduced by siting energy conversion facilities in areas with the most favorable conditions. Knowledge of technological and locational factors may be useful in reducing local or even regional impacts but does little to reduce the global impacts of certain energy strategies. Worldwide carbon dioxide increases from coal combustion are not addressed by adjusting technological or locational factors.

Coal Resource Transportation Systems

Transportation facilities are an important component of the coal resource delivery system. These facilities include railroads and slurry pipelines to handle coal, pipelines to transport synthetic gas and liquids, and high-voltage transmission lines for the transportation of electricity. Other modes of moving coal include barge and truck transport. Since coal conversion facilities often serve large, distant markets, the siting and construction of transportation systems are as essential as facility siting. The problems and issues involved in siting these facilities extend beyond the scope of this book. However, we should emphasize that problems related to transportation are another constraint to increased coal utilization.

Coal conversion facilities can be located at the mine mouth, adjacent to the market, or at some intermediate location. The choice of site depends upon policy and regulatory factors as much as on industrial location economics. Mine mouth facilities require the transport of converted energy to a distant market. Locating the facility near the market requires, in most cases, significant coal haul from the mines. Coal is currently transported by three major modes: truck, rail, and barge. Trucks provide collection and distribution services such as moving coal from mines to docks or local power plants. Rail and barge are long-haul carriers where available. Slurry pipelines can also be used as long-haul carriers for large point-to-point shipments. Railways transport almost two-thirds of all coal; highways, 12 percent; and barges, 10 percent (President's Commission on Coal 1980:194).

The most challenging problem involves long distance rail transport. The President's Commission on Coal (1980:200) reports that a $10 billion investment in the nation's rail system is necessary over the next eight years to meet anticipated needs. Major impacts of unit trains (trains which exclusively haul coal) are already being felt in many western communities which are divided in half as 100-car, slow-moving trains pass through (train speed limits are as low as 10 mph in some communities; U.S. Congress, Senate Committee on Energy and Natural Resources 1977). Slurry pipelines face severe political and environmental problems (Office of Technology Assessment 1978). These include the volatile issues relating to the use of western water to transport coal to other regions. Long distance electricity transmission causes significant energy losses and sometimes results in local impacts (Miller and Kaufman 1978). Young (1973) records instances of severe shocks occurring to individuals working on equipment in range of the electric field of a high-voltage transmission line. In response to growing protests over the siting of these facilities, many utilities have accelerated plans to build additional lines (*Business Week* 1977:27).

Siting and the Distribution of Impacts

The pattern of energy facility location is important in determining the type or level of impacts. Siting facilities at the mine mouth results in a different distribution of effects than shipping the coal to load centers for conversion and energy distribution. Mine-mouth siting, as its name implies, involves converting coal to a more usable energy form at or near the mine site and transporting the energy product to serve demand elsewhere. The proposed coal gasification facility in western North Dakota will convert local lignite into high-Btu gas which will be transported by pipeline to the Midwest. The siting strategy follows classical least-cost industrial location analysis where the conversion operation is a "weight-reducing" activity and the industry is "material-oriented."

In terms of the distribution of impacts, the local North Dakota area will receive tax benefits as the process adds value to the resource, and the state will receive additional revenues because of its high severance tax collections. The facility will also employ about 600 workers continuously, providing a source of stable employment and wages. On the other hand, the mine-mouth location results in the local area being subjected to undesirable impacts, while midwestern consumers receive clean gas with no environmental costs.

Most of the total air emissions resulting from energy development originate at conversion facilities, not at the mines. A "strip and ship" operation would transfer most of the air pollution to the region where the energy is to be consumed. Water requirements at the site of the resource are also less for "strip and ship" than for mine-mouth siting. *Energy from the West* (White et al. 1977:33) reported that water requirements for mining and reclamation are an order of magnitude less than that resulting from mine-mouth conversion facilities.

Other major categories of costs and benefits result from the population increases necessary to construct energy facilities. The ratio between the total number of construction workers and the number of workers needed for continued operation of the facility is important in determining the magnitude of "boom and bust" impact. Coal mines do not require as many workers as do conversion facilities during their construction stage. The peak employment for a coal gasification facility is estimated to be over twenty times larger than the peak employment for a surface coal mine (Carasso, et al. 1975:6-30). Because conversion facilities require so many more workers than most extraction facilities, the population-related impacts of mine-mouth siting are large. Housing problems, the provision of local services, growth management, and recreational issues occur as the pressure of a growing population is inadequately handled by small, isolated towns. Incoming populations are likely to disperse into remote areas for domestic and recreational activities, increasing ecological impacts by modifying wildlife habitat and contributing to illegal hunting and fishing. In western states, "the smaller impacts of mine construction and operation would not cause the social disruption predicted to accompany mine-mouth electric generation" (Metzer and Stenehjem 1977:8).

Large urban areas, such as Chicago or Los Angeles, with large numbers of skilled construction workers, would be better equipped to provide the manpower and services for constructing and operating new energy conversion facilities than would Gillette, Wyoming, or Beulah, North Dakota. It has been shown that a load-center conversion facility location would incur minimal population impacts as only a few people with special skills would be needed to move into the area compared to the massive immigration expected in the rural West (Metzer and Stenehjem 197:8).

A redistribution of impacts would occur when the negative aspects of conversion facilities are felt outside the resource region. Emissions from conversion facilities located at the load center would further degrade air quality in Chicago or Los Angeles rather than in small western towns. A qualitative distinction in air pollution issues would result from such a siting change. Whereas air pollution problems in the West are more of an aesthetic problem (the violation of PSD), ambient air pollution violations in the urban centers will contribute to a more critical health problem.

Siting and the Distribution of Impacts   19

The shortage of space near urban areas makes load-center sites less attractive. Waste disposal sites and storage space are difficult and costly to obtain in congested urban areas. Finding adequate land to dispose of energy wastes is a growing problem as many urbanized areas are already facing a shortage of space to dispose of municipal waste. Each day, the New York and northern New Jersey urban areas produce in excess of 28,000 tons of municipal wastes. The amount of available land for this purpose in the region has dropped from 2,500 areas to about 500 acres in a 10 year period (Committee on Science and Technology 1979:42).

Large coal conversion facilities require massive material, labor, fuel, and capital inputs and generate a host of residuals that are converted into impacts. While some of the impacts of large coal conversion facilities are viewed favorably (tax revenues), all facilities result in some serious undesirable impacts. Adjusting technologies or locations is a strategy to reduce or redistribute the impacts of energy facilities. Because of their potential undesirable impacts, coal conversion facilities are subject to a range of design and locational controls at the federal, state, and local level. Energy developers must secure permits and authorizations in order to site and construct these facilities. The extent of this control has been a matter of dispute. Many energy developers claim that siting procedures are too cumbersome and redundant. Others argue that there is insufficient control over facility siting questions. The next chapter provides an overview of this permitting process and ways in which it varies from state to state.

3

Permits and Approvals

A centralized coal conversion facility requires permits from agencies and offices at the federal, state, local, and sometimes regional level. A series of public meetings must also be held on the proposed project. Energy developers claim that there is a lack of coordination among these various agencies' requirements, redundancy in the permitting procedure, overwhelming and time consuming paperwork, and hearings causing additional delays. These concerns were voiced in the case of the Kaiparowits power project. If interest groups challenge a proposed project in the courts, the siting process enters a new realm of complexity. On the other hand, many analysts feel that the permitting process does not effectively protect human health nor the environment. It has been argued that public hearings do not provide a realistic avenue for public participation in energy projects.

Facility Siting Problems

Utilities and other energy companies engage in system planning to assess their future needs. The early stages of planning for new facilities are aspatial in nature. The utility decides that it must increase system capacity by building additional facilities. Important aspects of system planning that influence siting decisions for utilities are load forecasting, generator selection, reliability analysis, territorial considerations, corporate policy, and economics (Table 6; Cirillo, *et al.* 1976:5). Ordinarily, government agencies and interest groups are not active in the early planning stages for new energy facilities in most states.

Once the need for new energy facilities has been established, and assuming the utility decides to locate facilities on new sites, a screening process usually identifies several sites for further evaluation. The most significant criteria in evaluating the specific sites are engineering, safety, environmental, institutional, and economic considerations. Although the utility's system planning and site selection process must consider how the proposal conforms to federal, state, and local requirements, the utility's plans are largely proprietary and not subject to public inspection. Historically, once the utility persuaded the state utility commission and the Federal Power Commission (FPC) that a new facility was needed and that the project was economically sound, there was little public debate in the siting process itself. The utility would apply to the state utility regulatory commission for a certificate of public convenience and necessity. If this application were approved, it would acquire the site either by direct purchase or by

TABLE 6   IMPORTANT SITING CRITERIA FOR UTILITIES

CRITERIA	DESCRIPTION
System Planning	
Load Forecasting	Estimates of system demand (the need for electricity at any point in time) and the geographical distribution of the load.
Generator Selection	Choice of energy source (fossil, nuclear, hydro) and plant size.
Reliability Analysis	Study of the impact of plant location, size, type, transmission interconnections, and timing on system stability.
Territorial Considerations	Defining the region of interest and candidate areas for plant location.
Corporate Policy	For example, share of capacity met by each energy source.
Economics	Fiscal and other economic inputs to corporate decisions; for example, capital availability and cost.
Site-Specific Evaluations	
Engineering	Availability of adequate large-scalend area; sufficient cooling water, construction materials, and labor; suitability of foundation conditions; favorability of topography; accessibility of transportation facilities; and general plant and transmission line layout requirements.
Safety	Effects of accidents on the surrounding area and effects of the location and risks of accidents (*e.g.,* earthquakes).
Environmental	Impacts of a site on the physical environment, land use, regional development, and socioeconomic patterns.
Institutional	Regulations applicable for the area in which the plant is being located.
Economic	Comparisons among alternative technologies and sites in terms of capital costs, operating costs, and rate of return.

Source: Cirillo *et al.* 1976:8-14.

using its right of eminent domain and then build the facility after securing the necessary permits. If citizens requested information on the proposal, their attempts would be frustrated — they would find it very difficult to identify the individuals who did know the exact details of the project. This form of "purposeful ambiguity," as shown by Seley and Wolpert (1974), is a strategy that can diffuse public opposition by failing to provide sufficient information to challenge a proposal and by not identifying any individual or party that should be challenged. However, a number of changes have occurred during the past decade to transform this siting process into one of the most controversial aspects of domestic energy development.

First, as environmental concerns have increased during the past decade, numerous environmental laws and regulations which affect the siting of energy facilities have been enacted at federal and state levels. Individual citizens, organized interest groups, and governmental agencies are using these legal avenues to participate in siting decisions. Second, interest groups have successfully challenged siting plans and delayed final siting decisions for some facilities through participation in public hearings and use of litigation. Third, increased reliance on domestic energy resources and exclusionary implications of environmental legislation have forced many utilities to site in areas outside the one they serve. This pattern of development, most pronounced in the western states, has generated concern about regional exploitation and neocolonialism (Lamm 1976; Plummer 1977). Finally, specific legislation is now in effect in several states concerning energy facility siting. These state laws have provided a new platform for debate over the necessity of new energy facilities, have increased interregional conflict in some cases, and have created concern about the role of state planning in a time of "national emergency."

The Federal Role

Although federal regulatory agencies have jurisdiction over the siting of hydroelectric facilities and nuclear power plants, no federal agency has sole responsibility for siting of coal conversion facilities. Energy developers must be granted permits from federal agencies, or from federally-approved state agencies, in order to begin or continue work on fossil projects. Responsibility for siting is spread among federal, state, local, and regional governments. After the 1965 Northeast power blackout, the Federal Power Commission established the National Electric Reliability Council to coordinate power supply and interregional connections so that blackouts could be avoided. Power plant siting bills have also been introduced into congress to provide the federal government with more authority over coal conversion facility siting. Most of these bills were considered excessive federal encroachment into state land use decisions, and none passed.

The most direct avenue of federal jurisdiction over siting decisions for large private fossil projects is through environmental legislation. Single-purpose laws and regulations that protect the common environment require federal permits for large projects such as power plants. The federal role became more direct with the passage of the National Environmental Policy Act of 1969. Although this legislation did not call for the study of power plants and other large federal projects *per se*, it did require that an environmental impact statement (EIS) be filed for all proposed projects which require federal action and which will significantly affect the human environment (Greenberg *et al.* 1978). Because of their size, input requirements, and residuals, new coal conversion facilities require at least one federal permit. This means that the "lead" federal agency granting a permit must prepare an EIS. The geographical implications of the EIS have been evaluated in much more detail by Greenberg, Anderson, and Page (1978) and need not be discussed here. Table 7 lists the number and types of *major* permits and approvals that were needed in order to begin construction and operation of the North Dakota coal gasification facility. Since the facility was to obtain water from a federal impoundment of the Missouri River, a water withdrawal permit was required from the U.S. Bureau of Reclamation. The Bureau of Reclamation became the lead federal agency and filed the environmental impact statement. In the case of the SRC-II coal liquefaction project proposed in West Virginia, the U.S. Department of Energy, the federal sponsor of the project, filed the EIS as the lead federal agency. Table 8 lists the major permits and approvals required for this facility.

The U.S. Army Corps of Engineers is often the lead federal agency in power plant siting projects in the eastern states. The Corps' responsibility for navigable waterways dates to the Rivers and Harbors Act of 1899. With authority to maintain commerce on navigable waterways, Corps permits are required to construct loading docks or intake pipes for power plants on such waterways. In addition, the Corps was given authority in 1975 to establish procedures and issue permits for waste discharges into such waterways (Winter and Conner 1978:47-48).

An important requirement of the EIS is consideration of alternatives to the proposed project. Most early EIS's did not seriously entertain other alternatives, but recently, writers of these reports have been more conscientious about evaluating serious alternatives to proposed projects. Several impact statements indicated that conservation

TABLE 7   MAJOR PERMITS REQUIRED FOR A COAL GASIFICATION FACILITY IN NORTH DAKOTA

AGENCIES	PERMIT AND/OR APPROVAL
Federal Agencies	
U.S. Army Corps of Engineers	Easement for Water Intake, Pipeline, and Access Road; Section 10 Permits for Water Intake and Pipeline Crossings of Major Streams; Section 404 Permits for Wetland Disturbance.
Environmental Protection Agency	New Source Performance and Air Quality Significant Deterioration Review, Deep Well Disposal Review
Federal Power Commission	Certificate of Public Convenience and Necessity
Federal Aeronautical Administration	Application for a Notice of Proposed Construction for Structures over Regulated Heights
U.S. Bureau of Reclamation	Water Service Contract, Environmental Impact Statement
North Dakota State Agencies	
Public Service Commission	Plant Certificate of Site Compatibility, Water Pipeline Certificate of Site Compatibility, Water Pipeline Transmission Facility Route Permit, Mining Plan
Department of Health	License for Radioactive Measuring Device Operations, Hazardous Waste Control Plan, Wells for Temporary Water Supply, Sewage Treatment Plant
Environmental Engineering Division	Permit to Construct (Air Pollution Control Permit) Permit to Operate (Air Pollution Control Permit)
Water Supply and Pollution Control Division	National Pollutant Discharge Elimination System Permit for Deep Well Disposal, Solid Waste Disposal Permit
State Highway Department	Rail Siding Crossing, Pipeline Construction on Highway Right-of-way
State Water Commission	Appropriation of Underground Water, North Dakota State Water Permit
Secretary of State	Certificate of Authority for Foreign Corporations to Transact Business
Unemployment Compensation, Division of Employment, Security Bureau	Application for Coverage by American National Gas Coal Gasification Company
Workman's Compensation Bureau	Covered by American Natural Gas Coal Gasification Company
Local Agencies	
Board of Commissioners, Mercer County	Petition for Access to County Roads, Petition for Vacating County Road and Closing Section Lines, Certificate of Zoning Compliance, Plant Site Rezoning, Conditional Use Permit
Soil Conservation District	Erosion and Sediment Control Plan

Source: U.S. Department of the Interior 1977:1-9, 1-10.

efforts to slow energy demand would be more desirable than completion a proposed energy project. The final EIS for the SRC-II facility considered eight oil supply alternatives to coal liquefaction, no action, and two alternative sites for the facility (Table 9). While the alternatives to coal liquefaction were only briefly discussed, detailed studies were made of the alternative sites for the facility.

Other important federal environmental legislation protecting the aquatic, atmospheric, and terrestrial environments requires permits for large coal conversion facilities. The Federal Water Pollution Control Act Amendments of 1972 gave the Environmental Protection Agency permitting authority over energy facilities that dis-

TABLE 8   MAJOR PERMITS REQUIRED FOR A COAL LIQUEFACTION FACILITY IN WEST VIRGINIA

AGENCY, PERMIT

Federal Agencies

Environmental Protection Agency (EPA)
1. Prevention of Significant Deterioration (PSD) permit
2. National Pollutant Discharge Elimination System (NPDES) permit for construction runoff water
3. NPDES permit for discharge from sewage treatment plant
4. NPDES permit for water intake back-flushing and any other plant operating discharge
5. Resource Conservation and Recovery Act (RCRA) permit
6. Spill Prevention Control & Counter-Measure (SPCC) Plan

Corps of Engineers (COE)
1. Section 10 and 404 permits for construction in a navigable river

Federal Aviation Administration (FAA)
1. Notice of Proposed Construction permit

West Virginia State Agencies

Air Pollution Control Commission
1. Permit to construct, modify, or relocate an air pollution source

Department of Natural Resource (DNR)
1. Water Pollution Control permit for construction runoff
2. Water Pollution Control permit for sewage treatment plant discharge
3. Water Pollution Control permit for plant discharge operations
4. Water Pollution Control permit for a landfill
5. Dam Certificates of Approval

Department of Health
1. Permit to construct sewage treatment plant
2. Permit to operate sewage treatment plant
3. Permit to construct potable water supply system
4. Permit to operate potable water supply system
5. Permit to construct a Class III landfill for construction wastes

Department of Highways
1. Permission to enter highway

Department of Mines
1. Permit to plug a gas well

Local Agencies

No county or city permits required

Source: U.S. Department of Energy 1981:1-88.

charge effluents. The 1977 Amendments to this Act provide that energy facilities use the "best available technology economically available" to alleviate effluent pollution. Regulations stemming from the 1972 Act also promulgated cooling towers as the "best available technology" to control thermal pollution. This has meant that most power plants constructed after 1975 have cooling towers. As observed by Reynolds (1980:371), this regulation increases the attractiveness of clustered power plant siting over dispersed siting. Once-through cooling systems, popular before passage of these regulations, required that facilities on waterways be distant from one another to minimize cumulative thermal pollution.

All large coal conversion facilities require air pollution control permits. Many states have federally approved air pollution control regulations which meet or exceed federal air pollution standards. The nation has been divided into Air Quality Control Regions which are designated according to air quality. Allowable increments of air quality

TABLE 9   ALTERNATIVE ACTIONS TO A COAL LIQUEFACTION FACILITY IN WEST VIRGINIA

Alternative Liquid Fuel Technologies:
Increased domestic oil production
Oil shale development
Enhanced oil recovery
Outer continental shelf petroleum
Tar sands and heavy oil
Biomass and heavy oil
Coal — oil mixture

Alternative Sites:
Equality, Kentucky
Ravenswood, West Virginia

Source: U.S. Department of Energy 1981.

degradation are only allowed in certain regions so long as the new facility uses the best available air pollution abatement equipment and its emissions do not exceed federal standards. Federal regulations require that all new power plants install scrubbers as the best available control technology. Many parts of the western states have been designated as Class I regions, severely restricting energy development (Calzonetti, Eckert, and Malecki 1980). The ability to secure air pollution permits for power plants has been a major influence in the pattern of energy facility siting. Since many cities have air pollution levels that exceed federal standards, it is difficult to construct new power plants at the major load centers without offsetting the new pollution source with other reductions in air emissions.

Federal control over the environment is becoming increasingly important in the siting of coal conversion facilities. The Resource Conservation and Recovery Act of 1976 designated two types of wastes, solid and hazardous. Hazardous wastes must be disposed of in a particularly stringent and expensive manner which would add appreciably to the price of a facility's energy. Those wastes designated as "solid" must still be discarded in an environmentally-sound manner but not nearly so strictly as hazardous wastes. Solid wastes from power plants (fly ash, bottom ash, slag) were being studied by the Environmental Protection Agency to determine their classification. Wastes from coal liquefaction facilities will be treated as hazardous. Solid waste disposal is a great concern for synthetic fuel facilities because of the large volumes generated and their potentially toxic characteristics. The Resource Conservation and Recovery Act does not allow hazardous wastes to be stored or disposed of at certain types of locations (floodplains, wetlands, close to residences). Thus, many prime waterway locations for these facilities are more expensive because the operator must transport the waste products off site to a safe disposal location (Calzonetti 1979).

Federal permits are also required in response to protection of wildlife and historically significant cultural landmarks. The Endangered Species Conservation Act of 1969 and the Endangered Species Act of 1973 provided federal protection to threatened plants and animals and their habitats. The recent experience with the delay of the Tellico Dam project in Tennessee because it threatened the snail darter, an endangered species, demonstrates the potential of this legislation. The protection of cultural and historical places comes under the auspices of the Historic Preservation Act of 1966. This is designed to mitigate or eliminate impacts of projects on cultural properties that are or may be placed in the National Register of Historic Places.

The federal role in siting decisions has also been legitimized by a series of Executive Orders that have caused federal departments to reevaluate federal loans and grants to projects that affect floodways, wetlands, or prime agricultural lands.

The State Role

All states regulate electric power plants and have control over synthetic fuel facility siting decisions. The state regulatory control over electric power includes setting the retail rates for electricity, intrastate power transmission, and intrastate power pooling arrangements. A "certificate of convenience and necessity" or some equivalent is issued by the state Public Utility Commission demonstrating that the state accepts the utility's demand forecasts and the effects that the construction of an additional facility will have on the retail rate structure.

Procedures for permitting new facilities became more complicated as a result of environmental legislation passed in the sixties and seventies. Implementation and enforcement of this legislation was spread throughout many state agencies culminating in a decentralized and overlapping permitting process. Developers found this process confusing, time-consuming, and redundant; the states found it to be costly. Winter and Conner (1978) found that 21 state agencies were typically concerned in these situations. Thirteen West Virginia state permits are required for the coal liquefaction facility in Morgantown, a project which was not subject to Public Utility Commission jurisdiction (Table 8). Partly in response to the intractability of this process, states began to pass laws to streamline the energy facility siting process, largely by coordinating primary siting authority through one state agency. In 1972, only five states had legislation of this type, but by 1977 specific siting legislation existed in 28 states, and an

TABLE 10   STATE POWER PLANT SITING LEGISLATION

STATES HAVING MINIMAL POWER PLANT SITING LEGISLATION		
Alabama	Michigan	South Dakota
Alaska	Mississippi	Tennessee
Colorado	Missouri	Texas
Delaware	North Carolina	Utah
Hawaii	Oklahoma	Virginia
Indiana	Pennsylvania	West Virginia
Louisana	Rhode Island	
STATES HAVING MODERATE POWER PLANT SITING LEGISLATION		
Arkansas	Kansas	New Jersey
Georgia	Kentucky	New Mexico
Idaho	Maine	North Dakota
Illinois	Nebraska	South Carolina
Iowa	Nevada	Vermont
STATES HAVING EXTENSIVE POWER PLANT SITING LEGISLATION		
Arizona	Massachusetts	Ohio
California	Minnesota	Oregon
Connecticut	Montana	Washington
Florida	New Hampshire	Wisconsin
Maryland	New York	Wyoming

Source: Winter and Conner 1978:29.

additional 13 states had proposed such legislation (Southern Interstate Nuclear Board 1976; Eckert 1977). Table 10 indicates those states with minimal, moderate, or extensive power plant siting legislation.

Some state siting laws extend beyond the objective of coordinated permitting to using these laws as a tool for energy location planning. Many western states were aware that rapid energy development could conflict seriously with existing economic interests, non-energy natural resources (water), public welfare, and state heritage. Some of these laws were passed in conjunction with state severance taxes that aimed to mitigate the undesirable impacts of energy development. The North Dakota siting law is an example of legislation designed to protect fragile areas from energy impacts, a step toward state energy land use planning (Table 11).

Most states with siting laws have a single administrative body which acts as the lead agency for local, state, and often federal oversight of an energy project (Arkansas, New Jersey, and Wisconsin are notable exceptions). Its decision to issue a "Certificate of Convenience and Necessity" or reject an application is usually final, preempting regional or municipal challenges. The only recourse for the developer is state judicial review if appropriate statutes governing the state's site overview procedures were not followed, resulting in a denial of the developer's right to due process. The courts are not allowed to rule on substantive findings of the state's siting body. On the other hand, the state agencies may use the courts to enforce compliance with the conditions of the

TABLE 11   NORTH DAKOTA EXCLUSION AND AVOIDANCE AREAS

EXCLUSION AREAS:
(a) Designated or registered: national parks; national historic sites and landmarks; national historic districts; national monuments; national wilderness areas; national wildlife areas; national wild, scenic, or recreational rivers; national wildlife refuges; and national grasslands.
(b) Designated or registered; state parks; state forests; state forest management lands; state historic sites; state monuments; state historical markers; state archaeological sites; state grasslands; state wild, scenic, or recreational rivers; state game refuges; state game management areas; and state nature preserves.
(c) County parks and recreational areas; municipal parks; parks owned or administered by other governmental subdivisions; hardwood draws; and enrolled woodlands.
(d) Areas critical to the lifestages of threatened or endangered animal or plant species.
(e) Areas where animal or plant species that are unique or rare to this state would be irreversibly damaged.
(f) Prime farm land and unique farm land, as defined by the Land Inventory and Monitoring Division of the Soil Conservation Service, United States Department of Agriculture.
(g) Irrigated land.

AVOIDANCE AREAS:
(a) Areas of historical, scenic, recreational, archaeological, or paleontological significance which are not designated as exclusion areas.
(b) Areas where surface drainage patterns and groundwater flow patterns will be adversely affected.
(c) Within the city limits of a city or the boundaries of a military installation.
(d) Areas within known floodplains as defined by the geographical boundaries of the 100 year flood.
(e) Areas that are geologically unstable.
(f) Woodlands and wetlands.

Source: North Dakota Energy Conversion and Transmission Facility Siting Act, North Dakota Century Code, Chapt. 49-22-10 (1978).

permit during construction and operation of the facility. States often allow citizen suits to ensure this compliance.

State siting laws generally provide the means for a greater exchange of information with the public than is typically the case in states without siting laws. The EIS review procedure provides for public hearings in all states. States with siting laws typically have strict public hearing schedules. These public hearings have several formats, but contents typically cover interveners' perceptions, reservations, or technical findings related to the proposed facility's impacts.

State siting boards also require that the developer submit more detailed and comprehensive information concerning the proposed energy facility than is normally required by state utility commissions. Energy demand forecasts are carefully scrutinized by some state siting boards to ensure that new facilities are for the public good. Public Utility Commissions generally require information on the location, type, size, and related infrastructure of a proposed facility; many siting boards require specific information on the sources of water and fuel and a statement describing potential environmental and socioeconomic impacts resulting from the facility. Over half of the states with siting laws require this information for several possible sites. In California the developer must select a primary site and two feasible alternatives which are presented and evaluated by the State Energy Commission.

Municipal and County Oversight

Municipalities can affect the location of energy facilities through local zoning ordinances, building codes, health and sanitation standards, and taxation policies. Taxes collected by a local community are a form of compensation for the undesirable impacts generated by a coal conversion facility. Through zoning mechanisms, the community has authority to determine the location of energy facilities within its jurisdiction. Smaller communities or rural areas, where new facilities are often proposed, are in reality unable to exert much influence over proposed large-scale development because of a lack of technical expertise, political power, or commitment. Even more frustrating from the standpoint of a municipality is when a large facility is constructed nearby but outside its jurisdiction on county property. The municipality is subjected to many undesirable impacts of the facility, but the revenues are collected by the county and may be distributed for other purposes. Municipalities and county government may influence certain aspects of the development, but are not influential in making the big decisions on whether to proceed with development plans. Local government is most effective when the county and city act in coordination, preventing the developer from playing one jurisdiction against the other. Several states (California, Florida, Idaho, Nebraska, Oregon, and Virginia) require municipal and/or county comprehensive land use plans. Of these states, several require implementation, including the promulgation of ordinances, establishment of planning commissions, and appropriate enforcement measures. In these states, siting laws normally take into account the local plan and will not permit a development that contradicts local plans and ordinances.

Licensing Delays

Many industry representatives maintain that siting delays are a major element of the nation's energy problem. The plethora of permits, authorizations, and public hearings are identified as a leading cause of delay. Conflicts among competing interests also are instrumental in causing delay. The Kaiparowits case highlighted this utility viewpoint. Various factors are responsible for causing siting delay, not all of which are related to red tape or conflicts. Materials delivery problems, for instance, have been shown to be a much more important cause of slowing construction schedules than is usually recognized. In addition, slower growth rates reduce some of the urgency to bring a facility on-stream in a given year. It should be remembered, however, that modern facilities facing long lead times are much larger than predecessor facilities that were approved and completed in a much shorter period of time. An eight-year lead time for a 3,000 megawatt facility may be more reasonable than a four-year lead time on an 800 megawatt facility.

The U.S. Department of Energy (1979) maintains records from projected power plant projects that list the causes of delay cited by the developer (Table 12). In the case of coal-fired power plants, it is obvious that problems associated with prolonged permitting procedures and legal challenges are overstated. "Natural disasters," a broad term which includes events such as construction accidents, and financial and economic problems faced by the utility, account for delay in over two-thirds of the cases. Financial and economic problems include failures to have rate increases passed, or general economic conditions that reduce electrical demand. Although a delay category was provided by the Department of Energy to cite load forecasting errors, no utility recognized this as a source of power project delay. By contrast, permitting procedure problems and legal challenges are a much more important source of delay for nuclear power plant projects. Again, natural disasters and financial or economic problems are also cited as important factors slowing nuclear projects.

Since licensing procedures and hearings provide access for public interest groups to become involved in the siting process, siting procedures have been attacked as an avenue for interveners to cause delay. The utility view is summarized by Ward (1979:61):

> The upshot of this system which encourages public participation but then only allows him the tactic of delay is, not surprisingly, interminable delay. The intervener ultimately loses his case – but succeeds in significantly increasing the cost of the facility. Several years of hearings are not uncommon.

Federal and state legislation have provided avenues for increased public participation that were formerly available only through the use of common law. Common law, with its restrictive rule of standing, did produce notable citizen victories, but called for a

TABLE 12   CAUSES OF POWER PLANT DELAY

REASONS CITED	PERCENT OF CASES	
	COAL-FIRED	NUCLEAR
Natural Disaster	45.6	23.1
Financial or Economic Problems	24.6	16.1
Prolonged Procedures to Obtain Necessary Certificates from Government Agencies	12.2	25.2
Legal Challenges	6.1	14.7
Equipment Problems or Late Delivery of Equipment	4.4	4.9
Rescheduling of Associated Facility; e.g., Transmission Lines	3.5	1.4
Labor Problems	1.8	3.5
Changes in Regulatory Requirement	0.9	7.0
Strikes	0.9	4.2

Source: U.S. Department of Energy, Energy Information Administration 1979: Table 3

sophisticated organization with the knowledge, skills, and time to participate in lengthy hearings to attack and defeat utility projects. Standing refers to the right of an individual to use the courts. Generally, the rule of standing requires that only individuals who are able to prove that an action will damage his property or proprietary interests may take his case to the courts (O'Riordan 1976: 271-282). Class action suits, in which a group of individuals sharing a common interest may enjoy legal standing, have been used by many groups to challenge power projects. A collection of conservationist groups and individuals, known as the Scenic Hudson Preservation Conference, was granted class standing since the group (numbering 18,000) shared a common interest in protecting the Hudson River Valley from a hydroelectric-pumped storage project (Caldwell, Hayes, and MacWhirter 1976:218-227). This group was able to force the Federal Power Commission to modify the plans of the developer (Consolidated Edison of New York) and to consider environmental factors in its decision.

The permitting process is substantially lengthened by legal proceedings. Consolidated Edison began planning the Storm King Mountain hydroelectric plant in 1960 and envisioned that the facility would be in operation by 1968 (Caldwell, Hayes, and MacWhirter 1976:219-220). It was not until 1973 that the Scenic Hudson Preservation Conference lost its court battle against the New York Commissioner of the Department of Environmental Conservation.

Lack of Coordination

Common to the facility siting process is the overall lack of coordinated and comprehensive energy planning. Although state siting agencies and regional planning commissions are beginning to coordinate energy decisions, it is still true that new energy's place is being decided on a case-by-case basis, with little discussion of regional or national implications. The cumulative result of individual siting and permitting decisions (many made by single-purpose agencies) may lock the nation into a pattern of energy supply that will persist for many decades. Implications of shifting a large portion of energy supply to new coal facilities may be profound and unanticipated.

Although efforts have been made at both federal and state levels to streamline the siting process and reduce the overlaps and inconsistencies in facility permitting procedures, the process is still complicated. Overall, policy is made more on the basis of state and local interests than on regional or national needs. State and local agencies sometimes have difficulty regulating the activities of utilities and corporations that plan at the multistate level. While state siting commissions and environmental impact statements do consider the broader implications of a new facility, they generally focus on the impacts of the new facility on the surrounding area and do not evaluate regional or national siting strategies. State siting commissions give little attention to implications of facilities for interstate power systems. A notable exception is the Montana Major Facility Siting Act that lists the relationship of energy facilities to the regional grid distribution system as one criterion for siting. Because major electrical production facilities are tied together in a regional power pool, siting new facilities affects the reliability and efficiency of the entire system.

4

Centralized or Decentralized Energy?

When any new energy facility is proposed, disputes often arise over why it is needed. One reason that energy facility siting hearings are so acrimonious is that they become a forum for discussing energy policy, not the appropriateness of a particular response to an energy problem. A proposal to site a new power plant or other facility is challenged by those who favor energy conservation over energy supply solutions, those who advocate using other energy supply alternatives, those who would rather have the facility sited elsewhere, and those who are opposed to the technology. The discussion resulting from an energy proposal often becomes very broad because most states (and even the nation) lack a detailed energy plan that calls for an appropriate mix of energy supply and conservation alternatives. Susskind and Cassella (1980:17) found that this was a significant reason that siting discussions become so embroiled in larger policy issues:

> Without such a policy that enumerates production and conservation objectives, disagreements over the desirability of alternative energy sources or the relative desirability of alternative technologies will be played out every time a new project is proposed.

This contributes to the length of licensing procedures and the difficulties encountered in resolving conflicts over specific energy projects. Disputes arise when different interests disagree on demand growth rates and the need to increase system capacity. One guidebook for citizen activism against power projects (Morgan and Jerabek 1974:75-76) recommends that interest groups challenge utility projects on several counts to dispute the need to construct new energy facilities. The guidebook points out that appliance saturation can offset previously high energy demand growth rates and that utility policy can significantly affect demand.

Some siting conflicts arise when coal conversion facilities are proposed that many people do not believe will serve an important need. Many groups and individuals maintain that these facilities are pushed by institutions or authorities without a clear justification of their usefulness. Siting disputes often become entangled in the larger ideological controversies of "growth vs. no-growth" and "technocentrism vs. ecocentrism," making concensus on a particular project difficult.

The distinction between centralized and decentralized energy development strategies lies at the heart of many siting conflicts. In his classic 1976 article, Lovins clarifies this distinction:

> The first path . . . relies on rapid expansion of centralized high technologies to increase supplies of energy, especially in the form of electricity. The second path combines a prompt and serious commitment to efficient use of energy, rapid development of renewable energy sources matched in scale and in energy quality to end-use needs, and special transitional fossil-fuel technologies . . . (Lovins 1976:65).

Figure 4 illustrates some of the decentralized and centralized energy alternatives that can help to meet the goal of increasing the nation's energy supplies. Centralized systems include the familiar nuclear and coal facilities that supply energy to a regional or national distributional network (electrical transmission grid or gas pipeline network). Some centralized facilities can be based on renewable or continuous energy. The U.S. Department of Energy's 10 MWe "Tower of Power" solar thermal generating plant near Barstow, California, is an example of centralized energy facility based on a flow resource. Decentralized alternatives include residential systems (solar heating), community systems (district heating), and industrial systems (co-generation). Small, decentralized facilities may be based on non-renewable, renewable, or continuous resources. Commoner (1979:60) discusses a co-generator unit developed by Fiat known as TOTEM which converts 66 percent of the fuel's energy (gasoline, methane, or alcohol) into heat and 26 percent into electricity. Only 8 percent of the unit's energy is wasted (compared to over 60 percent in a large power plant).

Decentralized Energy

Decentralized energy systems are popularly known as "soft" technologies, a term introduced by Lovins (1976), or "appropriate technologies" as discussed by Schumacher (1973), in that the energy provided is matched to serve a particular local energy need. Love (1977:78) defines appropriate technology as "locally produced, labor-intensive to operate, decentralizing, repairable, fueled by renewable energy, ecologically sound, and community building." A goal of these systems is to reduce waste as much as possible. By definition, these systems are not designed to serve a wider market, although energy produced from them could be accepted into a regional power grid. It could be argued that where local energy demand is extremely high, such as urbanized areas, the appropriate energy technology is centralized power (such as nuclear or coal-fired power plants). However, advocates of the soft energy path would argue that new centralized facilities should only be used after intensive efforts toward conservation, co-generation, and district heating.

Decentralized energy systems, while not totally appropriate for urban areas, may be extremely valuable in rural areas. A commitment to low-head hydroelectric, solar, biomass, or wind energy may help to make rural areas almost totally energy self-supporting for heating and electricity. Reducing the reliance on centralized energy would keep money in the area instead of sending it to power companies. As the cost of energy increases, such transfers of wealth from rural areas may become increasingly significant. This concern has been raised by Messing et al. (1979:40):

> There is a fear that an "all electric" rural America will be a very expensive proposition for farmers and rural communities when all costs and externalities are included, particularly given inflation, rising construction costs, and rising energy costs . . . Large centralized systems, with

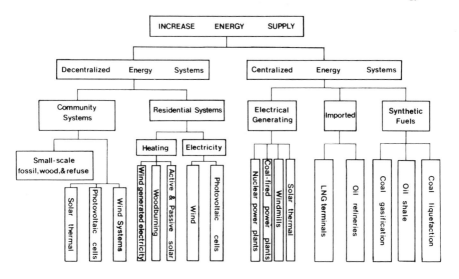

FIGURE 4 CENTRALIZED AND DECENTRALIZED ENERGY ALTERNATIVES

larger infrastructure costs and commitments (transmission and distribution) needed to service rural areas, may make rural consumers "high-cost dependents" and, perhaps, most vulnerable to higher rates during normal periods of service and most vulnerable to cutbacks in times of shortages.

The extent to which decentralized energy can help to maintain the economic strength of rural areas needs further investigation.

Decentralized facilities do not generate the siting problems associated with large centralized facilities. Since these facilities are much smaller, the lead time for constructing and having a facility operating is much shorter. Many federal and state laws exempt small facilities from the need to secure permits and authorizations. Thus, decentralized facilities can be deployed very quickly if the necessary inducements exist to spur individuals to adopt these systems. The National Research Council (1979:347) envisions this occurring with decentralized solar systems if the government takes strong measures to increase their attractiveness.

Another reason the siting of these facilities is not a problem is that most systems are relatively benign. In some cases, this is a function of the smaller scale of the facility, such as a windmill, that only affects a few people in a mildly disagreeable manner (interference with TV reception). In some solar space heating and hot water systems the only undesirable attribute of the system is the problem of disposal of transfer fluids (Weeter and Carson, 1979). On the other hand, some communities are beginning to experience serious air pollution problems as many residents adopt wood and coal stoves to fight rising heating bills. Decentralized technologies based on non-renewable and renewable resources may not be viewed as favorably as those using continuous energy forms.

A third reason why decentralized systems do not face the siting problems of centralized facilities is that they are constructed to serve local inhabitants and provide

FIGURE 5 LOCATION OF ANNOUNCED COAL CONVERSION FACILITIES (Benson and Doyle 1978; Rich 1978)

some measure of local or individual self-sufficiency. Success has been noted in several cities using an initiative to solve local energy problems through conservation measures and decentralized energy forms (Brunner 1980).

Decentralized technologies are less amenable than centralized technologies to state or national energy planning. Energy planners can make a few decisions on large facilities to increase energy supply by several thousand megawatts, whereas if they relied on decentralized technologies, an equivalent energy contribution could require millions of individual decisions. The federal government must disaggregate national energy goals to local areas and rely upon a system of incentives to motivate individuals and communities to adopt decentralized systems (Brunner 1980:85).

Centralized Energy

Centralized energy facilities provide power to a national or regional energy distribution system. While it would be impossible to map the location of future decentralized energy technologies, a map of some accuracy can be drafted to illustrate the location of centralized facilities due to come on-line in the future (Figure 5). Many announced facilities will probably not be constructed; they are on the long-term planning horizon of utilities and other energy companies.

Essentially, new energy facilities are ordered for three reasons: (a) to replace obsolete facilities; (b) to expand system capacity, and (c) to substitute new supply systems as other energy supplies are depleted. Centralized facilities may be ordered by energy utilities, private energy corporations, government entities, or a group of different energy organizations. The rationale for building new facilities depends upon the type of organization involved. Although a utility must decommission some facilities as they become obsolete, the motivation for ordering new facilities is usually to increase system capacity or to switch to an alternative boiler fuel. Large, modern power plants dwarf the capacity of predecessor plants so the replacement of obsolete facilities is usually only a contributing factor in siting questions.

The largest coal-fired power plant in 1955 had a generating capacity of 300 MWe, one-tenth of the capacity of recently announced 3,000 MWe facilities (Ford 1980:25). Figure 6 illustrates the trend toward larger facilities through time. An increasing proportion of our electrical generating capacity is represented by recent construction (Table 13). Forty-seven percent of existing capacity has been added since 1970. Projected coal-fired capacity will add 154 gigawatts to the existing 217 gigawatts, an increase of over 40 percent. The size of units was able to increase in response to design and engineering modifications developed after 1930, but power plant units have apparently reached optimal sizes in the 500-600 MWe range, and the construction of larger units is not anticipated. However, modern facility sites may have capacity exceeding 3000 MWe by having multiple generating units. Thus, the outcome of a few siting decisions could be important in determining how a region receives its principal electricity supply and what regions are the principal energy suppliers.

The decision made by an electrical utility to expand capacity is based upon forecasts for demand and the ability of the system to provide reliable service at a reasonable cost (Maher 1977:190-191). It is important for an electrical utility to maintain reliable electrical service to its customers. Understanding a utility's motives for wishing to site a new electrical generating facility requires some knowledge of reserve margins and mixes of base-load and peak-load facilities. Siting disputes are often complicated discussions of base-load or peak-load facilities, power sales or purchases, and appropriate reserve capacity levels. This complexity stems from the fact that electricity cannot readily be stored but must be produced in sufficient quantities to meet current

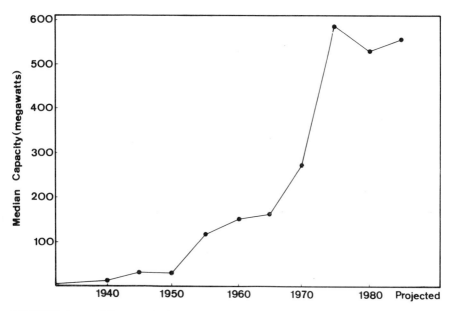

FIGURE 6   MEDIAN CAPACITY OF COAL-FIRED POWER PLANT UNITS (U.S. Department of Energy, Energy Information Administration 1979)

TABLE 13 U.S. COAL-FIRED ELECTRICAL GENERATING CAPACITY

CONSTRUCTION PERIOD	CAPACITY OF FACILITIES[a] (MWe)	PERCENT OF TOTAL
Before 1930	460	0.198
1930-1934	200	0.192
1935-1939	566.8	1.26
1940-1944	2,068.9	0.96
1945-1949	3,752.1	1.73
1950-1954	19,675.0	9.08
1955-1959	26,235.6	12.27
1960-1964	22,557.1	10.49
1965-1969	39,720.8	18.13
1970-1974	62.273.7	28.74
1975-1979	39,145.2	18.07
	216,655.1	100

[a]Facilities available for operation in 1979.
Source: U.S. Department of Energy, Energy Information Administration 1979.

demand. Since electricity demand varies seasonally and daily there are times at which a utility must be prepared to provide large quantities of electricity while at other times demand in the service area may be very low.

In response to this fluctuating demand, most electric utilities own "base load" and "peak load" generating plants. Base load facilities are designed to operate at close to maximum capacity almost continuously to provide basic electrical service to a utility's customers. When demand falls below base load output, electricity may be sold to adjacent utilities. Facilities with high fixed costs and low operating costs are often used as base-load facilities. As a rule, nuclear power plants are base load facilities.

When heavy demand is placed on a system, such as on a hot summer afternoon when air conditioners are operating, a utility will use additional facilities to provide extra generating capacity or purchase electricity from other utilities. Facilities that can be brought into service on short notice to meet high demand requirements are known as peak load units (Carlson, Freedman, and Scott 1979:11). Oil burning units, expensive to operate but easy to bring on-line, are often employed for peak-load situations. To meet peak-load demand, a utility may operate a pumped storage facility. A base load plant with lower operating costs will pump water into an elevated impoundment at times of slack demand. Turbines generate electricity during times of peak power demands.

Because a utility often experiences operating problems in its system or routine maintenance may close a plant, utilities must have a reserve margin so that even if some facilities are down, the system as a whole can still satisfy peak demand. It is customary that this reserve margin be 15 to 20 percent higher than peak load demands (Carlson, Freedman, and Scott 1979:8). The U.S. Department of Energy publishes reserve margin guidelines followed in different regions of the country. Utilities in a better geographical position to purchase power from other utilities need not have such a high reserve margin as those that are connected to only one or two systems. The practice of shifting electricity from one region to others through different utility systems to satisfy demand is known as "wheeling." This practice provides for more reliability in the national electrical supply system.

To coordinate electric utility companies in order to improve system reliability, the Federal Power Commission in coordination with Canada helped to organize the National Electric Reliability Council. The 1965 Northeast blackout was the impetus for this arrangement. Nine regional electric reliability councils were formed (Figure 7, page 38) which coordinate utility planning and data collection activities.

Utilities have facilitated electricity shifts by forming power pools to coordinate system planning, construction programs, and the buying and selling of electricity. Power pools operating at a multistate scale plan outside the scrutiny of state and local government agencies. The degree to which these institutional arrangements may limit public access to utility decision making has been identified as a serious concern by Messing, Freisema, and Morell (1979:53):

> Unless new institutional mechanisms are created to coordinate these planning functions, it would appear that local governments will remain unable to respond to planning options considered by regional utility planners and that an increasing amount of utility planning will be conducted through regional power pools or other interstate coordinating agreements with minimal consideration to options of local and even state governments.

Since centralized power is the cornerstone of utility development plans, important planning decisions in favor of large facilities and interstate transmission networks are being made without citizen input. This fact contributes to the distrust between different interests in siting hearings.

Although most Americans agree that new energy supply facilities need to be constructed, opinion differs widely on whether the emphasis should be on a decentralized or centralized approach. Part of the difficulty in reaching agreement on particular coal conversion siting decisions stems from the deep commitment of many individuals to decentralized power. Planning decisions for large facilities are often made at levels beyond the reach of private citizens, local officials, and, sometimes, state agencies, contributing to suspicions about centralized energy systems. However, it is clear that large coal conversion facilities will be constructed in the future. The next chapter provides more detail on the motives behind the siting decisions of principal coal conversion facility developers.

38 Centralized or Decentralized Energy

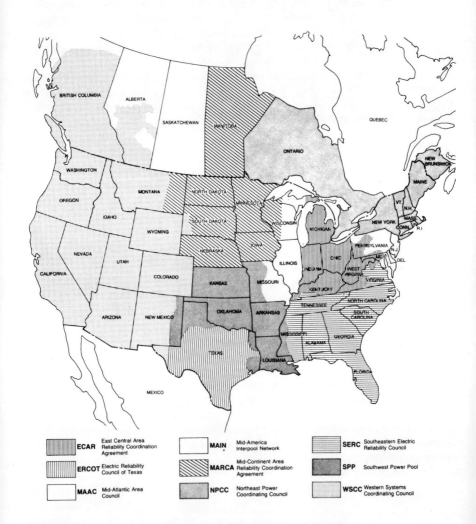

FIGURE 7   NATIONAL ELECTRIC RELIABILITY COUNCIL REGIONS (Source: NERC, *10th Annual Review*, August 1980. Reproduced by permission).

5

The Decision Environment in Energy Facility Siting

Geographers and regional economists have long been interested in problems relating to the siting of industrial firms. Is the process of selecting sites for energy facilities different than that for selecting sites for industrial facilities? Can the established analytical procedures used for industrial siting be used successfully for the siting of energy facilities? The goals and nature of decision-makers are important factors influencing both industrial and energy facility siting. The motives of energy developers, particularly energy utilities, are quite different from those upon which many industrial location models are based. In recent years, the value of location models for contemporary decision-making has been questioned along these lines. Thomas (1980:9-10), arguing that location theory does not accurately depict the motives of a modern organization, asserts that "classical location theory and its neoclassical economic foundations provide an inadequate framework for seeking coherent explanations for the industrial decisions of the firm." Many assumptions of classical location theory relating to the motives of decision-makers are even less able to provide adequate explanations for the location decisions of the organizations siting energy facilities. This means that industrial location models must be used with extreme caution when the topic concerns the siting of many different types of energy facilities. Given these caveats, industrial location models do serve an important purpose in energy facility siting. Variations of the Weber model, in particular, are used by many energy planners to help identify potential energy sites. In this chapter we indicate the conditions that must realistically be included in energy facility siting models. The next chapter provides a more detailed discussion of common siting approaches and alternative methods.

Location Theory and the Behavior of Large Organizations

Industrial location theory arose out of two major schools of thought, the "least-cost" approach and the "locational-interdependence" approach (Smith 1971). The least-cost approach, founded by Weber (1929), evaluated alternative industrial locations with respect to the aggregate cost of supplying the facility with its requisite inputs. As such, transportation costs were a prominent, and often predominant, factor in industrial location decisions. Markets were generally assumed to be punctiform, and little recognition was given to the nature of buyers or demand elasticities. In contrast, the locational-interdependence school, with its tradition from Hotelling (1929), shunned comparative cost analysis approaches in favor of an investigation of how spatial competition may affect the configuration of sellers to buyers.[a]

While there was some overlap in the two schools from the beginning, it is generally acknowledged that they were finally incorporated into an overall industrial location theory by Greenhut (1951-1952). These approaches have subsequently been refined and elaborated. The variable cost model, graphically described by Smith (1971), incorporated cost, demand, and dynamic conditions into a comprehensive framework.

Many researchers have attacked industrial location models on various fronts. Richardson (1969) disputed the assertion that profit maximization is the ultimate locational goal. This goal, as used in the variable cost model, leads to the choice of site which maximizes total revenues minus total costs. Richardson suggests that decisions within large corporations are based less on profit motives than on growth or "satisfying" behavior. Such conclusions have also been expressed in popular literature (Galbraith 1967; Scott and Hart 1979). Richardson notes that since the goal of a corporation may not be to maximize profits, the chances of constructing an operational model which incorporates their motives is severely limited. Richardson has been joined by numerous others calling for more detailed analysis of the motives of individuals within the firm to understand locational choice (Thwaites 1978).

Organizations Siting Coal Conversion Facilities

A number of different types of organizations are involved in siting coal conversion facilities. Except for privately-owned energy corporations (essentially integrated oil corporations) these organizations are either heavily regulated public or private utilities or government entities. The privately-owned energy corporation has the strongest resemblence to the "entrepreneur" considered in classical location theory but is only active in limited roles in a few synthetic fuel projects. Although these corporations own vast coal reserves, their activity in coal conversion facility siting is usually in concert with federal government agencies, utilities, or joint ventures with other corporations. Most coal conversion facility siting decisions involve power plants; the investor-owned utility is the major actor in this realm.

Privately-owned electric utilities generated over three-quarters of the nation's electricity in 1979 (Table 14). These corporations control almost 80 percent of the nation's generating capacity, and had total revenues of over $72 billion in 1979 (U.S. Department of Energy, Energy Information Administration 1980a:23). Privately-owned utilities operate 645 steam-electric plants with an installed capacity of 353 gigawatts.

TABLE 14 ELECTRIC UTILITY GENERATION IN THE U.S.[a] (Percent of Total)

YEAR	COOPERATIVES	FEDERAL	POWER DISTRICTS & STATE PROJECTS	MUNICIPAL	PRIVATELY OWNED[b]
1970	1.5	12.1	4.3	4.7	77.4
1975	1.9	11.5	4.8	4.3	77.5
1979	2.4	10.5	5.1	3.9	78.1

[a]Exclusive of energy used for pumped storage.
[b]Does not include industrial plant net generation.

Source: U.S. Department of Energy Energy Information Administration 1980a:15 1980a:15

Federally-owned systems account for slightly more than ten percent of the electricity generated in the United States. These systems include the Tennessee Valley Authority (TVA), the Bonneville Power Administration (BPA), and the Rural Electric Association (REA). TVA, a government-owned corporation, is the largest power system in the nation and provides electricity to about 2.5 million customers in seven states. TVA sells most of its power wholesale to local municipal and cooperative electric systems (Roberts and Bluhm 1981: 63-118). BPA markets power generated at federal facilities in the Pacific Northwest. Although it operates the nation's largest power supply network, it is not a major generator of electricity. The more than 1000 Rural Electric Cooperatives throughout the country own 40 percent of the nation's electric transmission lines. These organizations, originally formed to funnel federal funds into rural areas for electrification, play an important role in financing electrical power systems. Although they are not major generators of electricity, they are involved in major joint power projects with utilities since they are eligible for inexpensive federal loans (Messing, Friesema, and Morell 1979:28).

The Decision Environment

Power plant siting by these organizations is not based on free market principles, a fundamental tenet of classical location theory, but is carried out in a monopolistic — or oligopolistic in the case of non-utility energy firms — framework. The effect of monopoly and oligopoly operational environments is important and contrasts with the behavior of the firm as presented in classical industrial location theory. Energy utilities operate in a legal environment based upon their "mandate to serve" an area with a reliable energy supply. A particular utility is granted monopoly territory and a state-determined rate structure that allows cost-plus pricing with a regulated rate of return. Protection from competition with state control over rates and returns distinguishes utilities from true monopolies (Maher 1977:185). If a utility is unable to provide reliable service at a reasonable cost, competitive suppliers can be given access to its territory. Under this arrangement, utility executives are sensitive to the reliability of the energy supply system, even beyond profit maximization. Maher (1977:190-191), in a survey of midwestern electric utility executives, found that system reliability goals equaled or exceeded goals relating to the provision of service at a reasonable rate to the customer, and far exceeded their concern for covering costs or making an attractive return on investment. This does not imply that utilities are totally uninterested in holding costs down. As noted by Roberts and Bluhm (1981:52), state rate-making commissions have been under pressure to limit rate increases. Without sufficient rate increases, utilities may face problems securing their guaranteed returns on investments. This is one reason why many utilities are reluctant to equip power plants with expensive scrubbing units. There are some important implications of this monopolistic arrangement and of corporate goals for the siting of energy facilities.

Uniform Delivered Price

An important distinction between energy facilities and industrial facilities is the nature of demand. Most industrial operations produce marketable goods which are sold according to a demand function that reflects demand elasticities. Firms using f.o.b. pricing mechanisms allow the good's price to reflect the transport cost of supplying a particular customer. The product of a utility's energy facility is sold, by law, at the same price irrespective of location within a defined service area.

Naturally, the cost of generating electricity varies considerably throughout the country resulting in a wide range of electricity prices. National variation in electricity bills for large industrial customers has attracted energy intensive industries, such as aluminum smelting, to the Pacific Northwest because of its inexpensive electricity prices.

On the other hand, the cost of electricity does not vary within the service area of a particular utility. While inexpensive electricity may be an important regional factor in industrial site selection, it is not so important a site determinant within a region. A new energy facility does not necessarily attract energy intensive industries. Customers distant from the power generating facility pay the same price as those adjacent to the facility. The cost of providing service is certainly related to distance, making location relative to load centers an important consideration in the choice of energy site. If system reliability is the major objective and monopoly conditions exist within a cost-plus pricing system, it makes little sense to expend much effort to site facilities at profit-maximizing locations. For the most part, locational decision-making should parallel that of industrial facilities involved in an organized oligopoly which, according to Greenhut (1963:158), "does not promote an efficient distribution in space."

Energy utilities do not operate in a competitive situation. The regulatory environment defines the extent of the market (although incursions into it can be made by firms supplying alternative fuels), which demands a relatively predictable quantity of energy. The overriding preoccupation with reliability is manifested as a desire to control its operational environment so the "mandate to serve" can be fulfilled. To control its operational environment, a utility seeks to prepare for future contingencies by planning or influencing demand growth rates in its market area, assuring reliable fuel supply through long-term contracts, stockpiling fuels so that a particular facility can continue operation despite fuel delivery problems, and intensive lobbying activities to understand and influence the regulatory environment.

Capital Cost and Risk Minimization

Large coal conversion facilities are extremely capital-intensive. A large power plant costs in excess of $1.2 billion, and utilities report problems in raising money of such magnitude (Mitchell and Chatletz 1975). Gas utilities (currently changing their name to natural resource companies) face almost insurmountable problems in generating sufficient capital for synthetic gas facilities that will supply only a fraction of the gas provided by conventional sources. Because of the tremendous capital requirements of energy facilities a particular proposal is sensitive to project delays, which rapidly escalate the total cost of the project. The cost of the Kaiparowits proposal was rising at a rate of $1 million per day until the developers withdrew from the project.

Utilities and diversified energy corporations have responded to the problem of high risk from large financial capital requirements by pooling resources in consortium activity, enlisting the assistance of federal and state government, and even arranging financing with foreign governments. In 1975, American Natural Gas was the prime utility developer of the coal gasification facility in North Dakota. Unable to secure needed government loan guarantees, the utility pooled its resources with several other utilities to construct the facility. Rural Electric Cooperatives are often used by privately-owned utilities as a vehicle for securing low-cost Rural Electrification Administration capital and as a way to reduce financial risks in large power projects. Some 75 percent of Rural Electrification's loan guarantee commitments for 1975 financed cooperative participation with the electric utility industry for new power plants (Messing, Friesema, and Morell 1979:41).

Capital costs are spatially invariant. They only reflect location to the degree that the site will alter the construction costs of a facility. A remote location, for example, will require access roads, railroad spurs, and other infrastructural investments, increasing the total cost of building. Capital costs are incurred before the plant begins operation and is generating any revenue. Operational costs are more neatly incorporated in least-cost industrial location models. These include the costs for coal deliveries, labor, water, and any other costs that are incurred in the operation of the facility. Industrial location models, such as the variable-cost model, evaluate those factors that vary in cost with location. The cost of obtaining capital, for instance, does not vary with location.

High capital costs are most crucial for a utility trying to bring a new energy facility on-line. Each month of delay and each additional year that the plant does not generate revenue costs millions of dollars for most large projects. These costs become as significant as the locationally varying operating costs. A least-cost siting model may identify a site at which operation costs are low because of accessibility to the load center, the coal field, and transmission lines. These costs savings, however, will not be realized until after the plant is on-line and generating revenue, seven or perhaps ten years after the initial site survey was prepared. If the least-cost location incites interest group opposition, or requires additional hearings or other delays, financing charges escalate.

A "risk-reducing" siting strategy would be one that chooses a facility site that would meet state and federal laws and regulations and would not be likely to arouse citizen or environmentalist opposition. This strategy would allow the facility to come on-line as planned without excessive delay, thereby keeping capital costs at their lowest. If such a location incurs high operation costs, their expense will not be felt until far into the future, or may be balanced by capital cost savings. In addition, since the utility operates on a cost-plus basis, these costs can often be passed on to the consumer.

Least-Cost Siting Approaches

Least-cost location models, elaborations of the Weber model, are often employed to indicate candidate locations for new facilities. These methods are used to identify locations in which the major factor or factors affecting the cost of operation are minimized. Coal delivery and water availability are critical factors in the total operational costs of coal conversion facilities. Utility planners also seek locations that can be easily connected to the electrical transmission network.

Figure 8A illustrates a service area for a hypothetical electrical utility that has decided to build an additional coal-fired power plant. This utility provides electricity to one large city of a million inhabitants, several smaller centers, as well as a rural population. This electricity is generated at three power plants and is occasionally purchased from other producers during outages or to meet peak demand. For choosing a site for the additional facility, the utility uses a least-cost approach since revenues, as shown in Figure 8B, are not spatially variant. In this simplified example, proximity to coal, water, and the load center are the crucial costs varying spatially. Computing the cost of supplying prospective sites with these inputs provides a "cost surface" from which the final site will be selected. In this case, the utility chose a less than optimal site because of other conditions (availability of land). The utility had to locate within its "spatial margins to profitability" where total revenues equal total costs. It is doubtful that

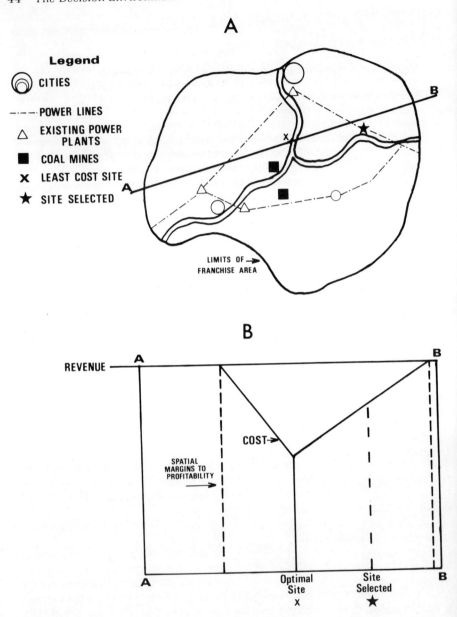

FIGURE 8   THE LEAST-COST SITING APPROACH

utilities undertake comprehensive searches to identify the particular least-cost location once they determine that their proposed site falls within the "spatial margins to profitability." This region, it must be stressed, is partly defined by the state cost-plus pricing structure. When the rate structure proves inadequate, the utility is forced again to seek rate increases from the state regulatory agency. This view of utilities' motives is not new or surprising. Labys, Paik, and Leibenthal (1979:19) succinctly summarized the nature of utilities' demands with respect to coal:

> *They are not too concerned about minimizing cost. In fact, their behavior reflects that management has preferred fuels more convenient and less expensive to handle than coal (such as gas and residual fuel) or simply more glamorous (such as nuclear fuels).*

Currently, electric utilities required by the Federal Power Plant and Industrial Fuels Use Act of 1978 (P.L. 95-620) to switch from oil and gas to coal as boiler fuel are meeting hostile receptions from utility commissions and citizen organizations who challenge their requests for rate increases (Shenon 1980). Rising costs and declining earnings, combined with high interest rates and inflation levels, have limited the ability of many utilities to undertake coal generator construction financing through such traditional means as bank loans and bond issues. Poor system planning and a belief that previously high growth levels would last indefinitely led most of the nation's utilities to invest heavily in large centralized facilities. Demand shortfalls in recent years and overinvestment caught most utilities short. In 1979, Standard and Poor's index of 22 electric utility stocks declined 10.7 percent to the lowest level since the 1975 recession (Shenon 1980:19).

Operational uncertainties can be reduced through site selection. In this regard, many utilities have an explicit policy of siting all new facilities within their market area. However, Figure 8 indicates that the spatial margins to profitability could extend beyond a utility's market area boundaries. If this is true, a utility may consider siting new power plants outside its market area. "Outsiting" trends have been considered by Hillsman and Alvic (1980), who found that:

> . . . *outsited generating capacity has increased from roughly 10 percent of the nation's capacity in 1950 to roughly 23 percent in 1980, and that it will probably increase to just under 30 percent in 1990.*

A utility could argue that no locale in its service area would accept the undesirable consequences of a new power plant, but it may be difficult to rationalize the siting of a noxious facility in an area that will not receive any major energy benefit. Control of mining operations and ownership of coal cars are strategies used by utilities which reduce uncertainty concerning fuel supply and leave the utility free to choose between "outsiting" and intramarket site.

The regulatory environment governing site selection can be a critical variable in further focusing the site decision. As an example, a location adjacent to available coal supplies could reduce total transport costs over the life of the facility by eliminating the costs associated with rail or slurry pipeline to a load center site. States that have shown a lack of cooperation in locating coal-fired generators that will serve out-of-state markets can make the siting process costly in terms of time and money. Such delays are clearly unacceptable to utilities that must provide for their market demand while staying within cost levels dictated by their rate structures, providing a favorable rate of return to investors, and presenting a picture of solvency to the financial establishment.

Traditional location theory would suggest facility site selection based on cost minimization or profit maximization. Instead, we see firms choosing sites based on the political realities of their operating environments. To be sure, economics are critical to the utilities' decision to expand capacity, but we are dealing with the economics of complex internal and external decision environments. Internally, there are existing plants, varying by capacity, age, fuel type, and system role — peak versus base load. These are being depreciated under different schedules and require varying amounts of labor and operation and maintenance costs, all of which are changing rapidly in time. In addition, before the decision to expand production capacity is made, other alternatives such as "wheeling" must be examined. Along with forecasts of future demand, the preceding criteria influence choices of fuel type and plant capacity. Site selection is a decision made later in the process. It appears that for utilities, optimizing the compatibility between the site and the energy technology is not a high priority. This is evidenced by the use of environmental impact statements as justification for site selection rather than as a planning tool and by the fact that the EIS often ignores the social, economic or political consequences of the siting decision. The result for the utility in these cases can be costly and long-lasting court battles that threaten system stability or, stated simply, threaten the utility's "mandate to serve."

In sum, we can see that the decision-making environment in siting is complex, and the motives of energy developers are quite remote from the "entrepreneur" assumed in traditional location theory. This does not negate the usefulness of locational analysis techniques in siting decisions, but calls for a closer approximation of the goals sought in finding sites for new facilities. In the next chapter we review a number of approaches that are currently used in finding sites for new facilities and other promising methods which attempt to operationalize multiple goals into useful siting tools.

6

Analyzing Siting Options

Various approaches to the problems involved in siting energy facilities have been developed. These include methods to identify and evaluate potential sites and techniques to resolve locational conflict problems. Geographers are active in both areas. Many geographers have helped to develop methods that indicate locations likely to be acceptable for new energy facilities. Generally, the geographer's role has been to evaluate locationally-varying siting factors in order to identify a few sites which can be inspected by engineers, geologists, and other technical professionals. Site screening methods and spatial allocation approaches are valuable tools in evaluating potential energy facility locations. Geographers also contribute to reducing locational conflict associated with siting, including various participatory planning strategies, compensation, and mitigation.

Site Screening Methods

The beginning point in finding a location for a new energy facility is to eliminate unacceptable locations so that a more serious evaluation of a few sites can be made. Exclusionary screening is a popular approach to narrow down the choice of site from an overwhelming number to a few serious possibilities that have a high likelihood of approval. Many utilities and consultants use map overlays to eliminate areas with unsatisfactory attributes. Water availability is often a crucial factor in the early screening stage (OECD 1977). Initial screening is usually applied for six considerations: (a) system planning, (b) safety, (c) engineering, (d) environment, (e) institutional constraints, and (f) economics (Cirillo et al. 1977:6-7).

The screening approach will identify candidate sites that are studied in much greater detail. Once a few proposed sites are identified, the utility may initiate contact with local authorities, planning departments, and landowners to begin on-site surveys, soil analysis, test drilling, and other evaluation methods to determine locational feasibility (OECD 1977:9). Many characteristics are examined in evaluating a particular site (Table 15). Once local officials are contacted, the utility may be made aware of other promising sites in the vicinity. If an otherwise attractive locale is not chosen for the planned facility, it is customary for a utility to put it into a reserve category (Cirillo et al. 1977:7).

Although exclusionary screening is a quick way to focus attention on a more manageable number of sites, it may eliminate some potentially good sites while mediocre sites remain. The screening process divides areas into acceptable and

TABLE 15   DESIRED ATTRIBUTES FOR ENERGY FACILITY LOCATION

Proximity to load centers	Long-term fuel supply available
Highway, rail, water, access	Effect of plant and transmitter line appearance on surrounding area
Good geological building foundation	
Good local weather conditions (infrequent strong storms, temperature inversions)	Amenities for employees
	Taxes
Good hydrological conditions	Sufficient land available for power plant, coal storage, waste products, loading/unloading facilities, parking
Unlikely location for floods	
Available cooling water	

Source: Energy Policy Staff 1968:7-16.

unacceptable locations on the basis of cutoff levels for *each* attribute. If any location is unacceptable for even one attribute, then it is eliminated from the study. If the attribute is a legal requirement, such as prohibition of a wetland location, then this procedure is sound. However, many attributes are discretionary and should not have rigid cutoff values (distance from water supply or load center). Hobbs (1980:189) calls for caution in the use of exclusionary screening methods:

> *Exclusionary screening is best used when there are legal and technical criteria that cannot be violated. If discretionary attributes are also considered, exclusionary cutoffs should be chosen with the realization that they imply tradeoffs and are arbitrary. Sensitivity analysis should be performed to see if locations are excluded that are otherwise superior.*

This point, illustrated in Figure 9A, is also emphasized by Keeney (1980:93). A siting study conducted for the Washington Public Power Supply System (WPPSS) eliminated areas farther than 10 miles or greater than 800 feet above a water supply using map overlay techniques (Keeney 1980:49-53). Keeney considered a situation in which three hypothetical sites (A, B, and C) are evaluated on the basis of these criteria. Sites B and C are eliminated because they exceed the cutoff value for one of the screening criteria. Site B, although adjacent to the water source, is just over 800 feet above it and is eliminated on the map overlay. Site C requires little vertical pumping but is located just over 10 miles from the water supply. Site A is acceptable since it is not more than 800 feet above or further than ten miles from the water supply. Clearly, site A is more expensive to provide water than sites B or C.

Keeney suggests that for such discretionary attributes, an effort at "compensatory" screening should be made. Figure 9B traces a line that represents water pumping costs of $7 million per year. All sites that incur pumping costs in excess of this amount are excluded by map overlay techniques. Thus, site A is eliminated from consideration while B and C remain feasible for further consideration.

Exclusionary screening has been refined to allow decisionmakers to make tradeoffs among the attributes. An energy developer may consider proximity to water more significant than good highway access, and may be willing to trade off one attribute for the other. Weighting summation is used to evaluate sites on the basis of attributes that have different levels of importance to different groups of decisionmakers. The optimal site is that which maximizes:

$$\text{Site suitability} = \sum_{i=1}^{n} w_i v_i(x_i)$$

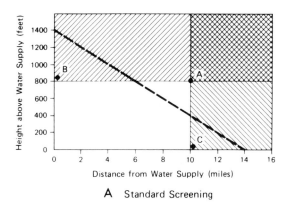

A Standard Screening

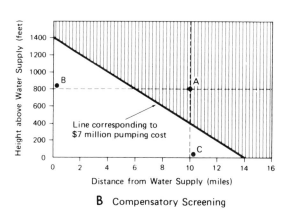

B Compensatory Screening

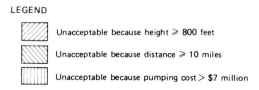

FIGURE 9 STANDARD AND COMPENSATORY SCREENING APPROACHES (Keeney 1980:93). Reproduced by permission of R.L. Keeney and Academic Press.

where $v_i(x_i)$ is the value function for attribute x_i, and w_i is the weight of the attribute (Rowe et al. 1979:11-12). Since sites are evaluated by summing the weighted value functions for each site, it is necessary that the attribute value functions be on an interval scale and the weight on the ratio scale. Hobbs (1980:189) cites cases incorrectly using the technique by weighting and summing ordinal value functions.

Dobson (1979) describes a weighted summation procedure used by the State of Maryland to identify those sites most suitable for new nuclear and coal-fired power plants. This case is important because of the comprehensiveness of the analysis and its public policy potential. The procedure required the identification of those attributes at a site (such as proximity to streamflow, endangered species, population density) that would affect a particular facility's costs and impacts. The relative importance of these variables for different decision-makers (utility executives, state planners) was assessed and the overall compatibility of a site considering all variables and their importance was determined. In the analysis, 31,234 cells of 91.8 acres each were evaluated across 52 different variables (Dobson 1979: 226). The siting objectives evaluated included the minimization of construction and operation costs, the minimization of adverse ecological impacts, the minimization of adverse socio-economic impacts, and a composite of all objectives. Dobson illustrated the outcome when the method was used according to the Maryland Power Plant Siting Program (Figure 10). In order to minimize construction and operating costs of a coal-fired power plant with cooling towers, the program staff ranked proximity to streamflow as the most important characteristic of a good site (Dobson 1979:229). The same group ranked endangered species and proximity to fish spawning and nursing areas as crucial variables when selecting sites on the basis of minimizing ecological impacts. Candidate areas were identified with regard to economic, ecological, and socio-economic factors. The composite map indicates areas that meet all three objectives. These can be interpreted as potential sites which meet the engineering and economic prerequisites of the utility while being less likely to cause conflicts because of potential ecological and socio-economic impacts. These areas can be evaluated in much more detail in a way similar to the final stage of the site-screening process.

This knowledge has been implemented by the State of Maryland in its energy facility site banking program. The state has been able to set aside locations for future energy facilities which have been determined to meet the requirements of different interest groups. When a new facility is needed, the utility chooses among these preselected sites. Most states lack an adequate geographic data base to undertake this type of analysis.

A similar screening method was also used for the California siting program which identified 61 significant "constraint" and "opportunity" factors that were mapped statewide at the 1:500,000 scale (California Energy Resources Conservation and Development Commission 1977:17-20). Constraining factors include those that would restrict siting or could be adversely affected by a nearby energy facility. An opportunity, on the other hand, would be a condition favorable to a new facility, an example being stable geological conditions. The resulting maps provide a basis for utilities to choose sites that are likely to be approved by the State Energy Commission. For instance, a utility could avoid scenic coastal areas or primitive areas while preferring navigable waterways which are viewed as favorable locations for energy facilities.

Oak Ridge National Laboratory has been digitizing a data base for the entire nation, including both locational and technological characteristics that can be employed for facility siting analysis. This can be used to determine trends in plant siting, constraints to siting in different regions, and implications of policy decisions affecting the distribution of new energy facilities.

Spatial Allocation Models

Spatial allocation models are useful tools for evaluating energy facility locations with respect to energy flows. Although the models are structured to examine commodity flows, their usefulness in analyzing facility sites has been repeatedly demonstrated in the geographical literature. As noted by Scott (1971), the models portray an ideal

CANDIDATE AREAS FOR A FOSSIL-FIRED POWER PLANT IN NORTHERN MARYLAND

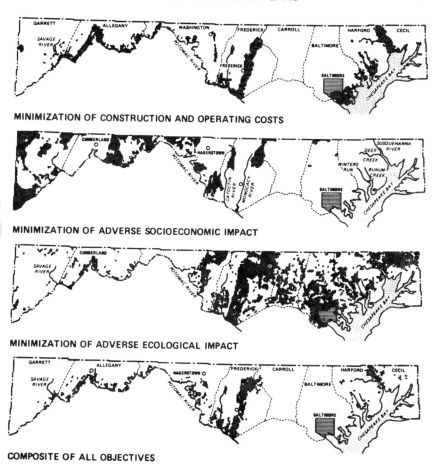

FIGURE 10 MARYLAND POWER PLANT SITE SCREENING. Reprinted from Jerome E. Dobson, "A Regional Screening Procedure for Land Use Suitability Analysis," *The Geographical Review*, Vol. 69 (1979), p. 227, with permission of the American Geographical Society.

system of flows from supply sources to demand destinations so that resource and transportation costs are minimized. The model can accommodate the existence of conversion facilities (power plants) that accept raw materials (coal) to be converted into a useable product (electricity) distributed to consumers. An evaluation of the model's output (the dual) reveals the comparative locational advantage of some facility sites or supply regions over others. This normative model can indicate how the supply system should be structured under ideal conditions, a useful tool in public and private planning (Chisholm 1971:130).

Osleeb and Sheskin constructed a model of the North American natural gas supply system to investigate potential future surplus and deficit natural gas regions. Their results indicated, among other things, that the Middle Atlantic, most of the Midwest, and South Atlantic states will likely be experiencing natural gas deficits after 1985 (Osleeb and Sheskin 1977:82-84). Government and industry planners can use such information in evaluating the possibility of locating coal gasification facilities in these areas.

A large number of linear programming models of the nation's energy supply systems were developed after the 1973 Arab Oil Embargo. The purpose of these models was to evaluate policy alternatives and estimate energy prices based on meeting growing energy demands with limited energy resources. The most ambitious and comprehensive of these models was the Project Independence Evaluation System (PIES) which was to outline strategies for meeting President Nixon's goal of attaining energy independence by 1980. As noted in a review of these energy models by Cohen and Costello (1975), some were aspatial in nature, focusing on economic sectors rather than regions, while others that offered more geographically valuable information could only provide limited assistance in facility siting because of their generalizations regarding transport modes, transport links, demand and supply areas, and facility sites.

Bechtel's Energy Supply Planning Model (Carasso et al. 1975) identifies the number of new energy facilities that must be constructed to meet a mix of energy demands for a future year. However, the model only locates the facilities in one of fourteen regions in the country, much too general for most needs. The model does allocate the fuel from the facility to meet U.S. demand and determines the requisite transportation facilities. However, as noted by Schanz, Sawyer, and Perry (1979:6-12), the model is less important as a siting tool than as an assistance to energy developers in evaluating the feasibility of various energy supply mixes in terms of the time, capital, manpower, materials, and construction schedules required for alternative energy supply systems. Even those models displaying the greatest spatial detail had transportation generalizations that weakened their overall usefulness for facility location planning, as was the case with the Battelle-EPA Energy Quality Model (Cohen and Costello 1975:23):

> The most appealing feature of the Battelle-EPA model is its spatial detail. The model can consider each of the 238 air quality control regions in the continental United States as an energy demand region. . . . Each demand region and supply district is designated by an x-y coordinate at its centroid and the crow-flight distance between them is used to estimate transportation costs.

Many early energy models, largely developed by economists and engineers, displayed sophisticated methodology, economics, and engineering but were naive in their treatment of spatial relationships and transportation. This is partly explained by the goals of the models and also the fact that the researchers had little geographical training, combined with the unavailability of energy data at a high spatial resolution. These models could not be very helpful to a public utility commission or other planning agency charged with evaluating energy sites since these national models, while indicating energy relationships between regions, did not identify intraregional implications.

More recently, the geographical sophistication of the models has caught up to their economic and methodological sophistication. The Brookhaven Regional Energy Facility Siting Model (Maier and Hobbs 1978) is one spatial allocation model that can be useful to energy planners. This multi-commodity, transshipment-location model is based on county-level data and can determine the least-cost distribution of facilities subject to environmental regulations, local environmental conditions, and technological factors (Maier and Hobbs 1978:3). Because of relative ease of formulation, the model can be useful to state energy planners and utility commissions who have primary responsibility for approving energy facilities and implementing siting plans but are handicapped by limited financial and manpower resources.

Spatial allocation methods can be a more powerful tool in planning the location of new energy facilities if linked to a regional screening approach. Once potential sites that meet various important criteria (e.g., economic, environmental, socio-economic) are identified, a spatial allocation approach can indicate which of these sites should be developed based on projected commodity and energy flows.

Resolving Locational Conflict

Once a candidate site for a new energy facility has been selected, the next problem involves bringing the facility on-line within a reasonable time period. This section briefly summarizes some approaches beyond those provided by state legislation that have been useful in expediting the siting decision. It must be remembered that quickly reaching a negative decision is also a measure of the success of the method since it saves time and money for all parties. Early knowledge that a candidate site is not acceptable will hasten the ability of the utility to consider alternative locations. Methods to streamline the permitting process as implemented by many states should not be accepted without a word of caution. As emphasized by Warren (1978), a system of multiple permits and redundant hearings and approvals, while an obvious nuisance to the energy developer, may be beneficial in guarding against hasty decisions or those made by individuals who may be oversympathetic to the developer.

Five approaches to conflict resolution have been identified in the literature (Thomas 1976:889-935; Blake and Mouton 1964; and Clark and Cummings, Jr. 1981). These are: (a) collaboration; (b) competition; (c) accommodation; (d) negotiation; and (e) avoidance (Table 16). The collaboration approach aims for a "win/win" outcome in which the goals of both parties are satisfied. Collaboration requires that both parties to a conflict share information and seek alternatives (Clark and Cummings, Jr. 1981). In the case of facility siting, it may be possible to adjust locations or technologies to meet the energy needs of the developer while satisfying the desires of other groups. In some cases, though, this is not possible and a "zero/sum" situation evolves, meaning that a gain to one group will result in a corresponding loss to another group. The two extreme cases are competitive and accommodation strategies. A competitive strategy results in

TABLE 16 OUTCOMES OF APPROACHES TO CONFLICT SITUATIONS

CONFLICT RESOLUTION APPROACH	OUTCOMES	
	MORE POWERFUL GROUP	LESS POWERFUL GROUP
Collaboration	WIN	WIN
Competition	WIN	LOSE
Accommodation	LOSE	WIN
Negotiation	COMPROMISE	
Avoidance	IMPASSE	

Source: Adapted from Gladmir and Walter 1978, Thomas 1976, and Clark and Cummings, Jr. 1981.

the group that has power meeting all of its objectives, ignoring the other group's wishes. An accommodation strategy is a situation in which the group with power takes an unassertive role and allows the other group to meet its objectives. A negotiation strategy is a compromise between the competitive "win/lose" and accommodative "lose/win" approach so that some of the objectives of each group are met while some of the objectives are not met. In an avoidance approach, both groups refuse to negotiate, resulting in an impasse. The conflict resolution literature is voluminous, but we can provide an illustration of promising strategies to reduce the conflict in siting coal conversion facilities.

Increased Citizen Involvement

As noted by White et al. (1977), three approaches are available to increase public participation in the energy facility siting process: (a) provide more information exchange between participants and agencies; (b) encourage administrative interaction between them; and (c) allow direct participant input into agency decisions. One essential ingredient of these approaches is the availability of reliable and credible information to all interested parties (Kash et al. 1976).

Public hearings are the most common vehicle for information exchange in facility siting decisions. Largely because of NEPA, every major energy decision requires a public hearing prior to licensing. Even in those rare instances where NEPA does not apply, a state agency, such as a state Air Pollution Control Commission, will have a hearing, although the range of topics discussed may be less comprehensive. At a hearing, it is customary for the developer to provide a brief description of the proposed project. Then individuals from the community make statements for the record voicing their concern or support. The agency presiding is to consider this information when deciding whether to approve or deny a permit.

Some doubts have been voiced over the value of hearings in encouraging maximum public participation. Many concerned individuals may not wish to voice their opinions openly in a formal setting in front of a large audience and the press. Hearings have also been criticized because of the failure of agencies to provide adequate notification. It has been suggested that a freer exchange of information could result by restructuring the hearings so that they take place in small, more informal settings (White et al. 1979:73).

The Council on Environmental Quality issued regulations in 1978 requiring agencies of the federal government to follow particular procedures in conducting an EIS process. It is now required that the lead agency must communicate with affected governmental agencies, the developer, and any interested group or individual to determine as early as possible the significant issues which need to be addressed in the EIS. Marcus (1981) recommends that an early hearing be conducted such that adversarial conflicts are minimized and consensus-building is maximized. One way to accomplish this is to use a "neutral facilitator," a specialist in meeting dynamics, whose role is to identify key issues and impacts to be addressed in the EIS and invent alternatives worthy of consideration (Marcus 1981:63). Such methods may allow negotiation or collaboration when different groups establish common ground on certain key issues. In cases where parties to a conflict are at an impasse, mediation may be helpful. This involves the use of a third party, who acts without authority to impose a settlement, who can assist parties toward defining their priorities so that compromise is possible (Carnduff 1981). Long used in labor disputes and international affairs, mediation is becoming an alternative in environmental disputes.

Another chronic problem voiced by concerned citizens is that the developer has not tried in good faith to communicate with them. The announcement that a major coke facility was planned for Morgantown, West Virginia, was only forthcoming after bulldozers started clearing the site. In addition, citizens complain that the press releases and impact statements are overly technical and vague. It has been recommended by White et al. (1979:730) that the developer be subject to a "participation audit" prior to the issuance of a permit to check if the developer has made a reasonable attempt to involve affected citizens and relevant agencies in the siting process. A permit would not be issued until the developer could demonstrate that ample consideration was given to the impacted population. The approach could also be helpful in facilitating communication early in the siting process.

The New England River Basins Commission (1980) in conjunction with the U.S. Geological Survey developed alternative site selection processes that increase citizen participation. These involve the interaction of the utilities, state government, and a task force of concerned groups. A goal is to involve the public in the decision process as early as is possible.

Citizen review boards have also been suggested as an approach to provide citizens with more direct access to the siting process. Citizen review boards would have power to make siting decisions rather than a solely advisory role. A review board would provide a forum for all interested parties and could act to reduce the levels of conflict that often prevail between different groups. The review board, as well as acting to deny a permit, could call for a redesign of the facility to reduce problems that are of particular concern to residents.

Naturally, if these mechanisms do not work, citizens still have the avenue of common law to influence siting decisions. Common law suits have been important in stopping many large projects, and could be a delaying factor in almost any case. Citizen suits are particularly powerful since "standing" would not be difficult to establish if a landowner's property will be affected by the proposed development. Since developers wish to avoid the courts to prevent the process from being overly lengthened, the threat of legal action by citizens and community groups is a powerful bargaining tool against a recalcitrant utility (Wolpert 1976).

To make public participation more effective, it is necessary to eliminate some of the manpower and financial handicaps that limit their abilities. Providing interest groups with funding or technical consultants would reduce the hardships imposed on a few individuals who must work part-time without financial compensation. The provision of attorney's fees, expert witness fees, and other reasonable costs with the aim of providing responsible input into a state or federal decision could greatly assist citizen groups and provide them with the staying power that is difficult to sustain without financial support (Metz 1977; Davis 1976).

Mitigation and Compensation

Impact mitigation refers to measures taken to alleviate the socioeconomic effects of energy development. Other energy development impacts, such as air and water pollution, can also be minimized, typically by technical adjustments in plant design rather than later mitigation. The socioeconomic effects of energy projects include additional demands on public and private services such as housing, streets, sewage treatment, and medical care. In small, isolated towns, these effects give rise to a series of social stresses commonly called "boomtown" problems (Energy Research and

Development Administration 1977; Gilmore and Duff 1975). Locational conflict often occurs because local residents are unwilling to tolerate these undesirable changes in their communities.

Mitigation strategies for these impacts, therefore, include financial assistance for towns which absorb large new populations without receiving additional taxes. In the western United States, many plants are located outside municipal boundaries and provide no direct tax benefit to the towns. Housing shortages are particularly acute, with mobile homes providing additional units. A large-scale attempt to mitigate community impacts has been successful in Colstrip, Montana, where the energy developer built a new town, complete with shopping and recreational facilities, to serve coal mine and power plant workers.

Housing and municipal finance impacts can be dealt with in three general ways which represent different interpretations of where responsibility lies. The federal or state government can make funds available for impact mitigation and effectively spread the cost among a large number of taxpayers. This is the approach embodied in the Coastal Zone Management Act of 1972 that provides aid to coastal zone energy areas and the 601 program (of the Power Plant and Industrial Fuel Use Act of 1978) for inland energy development administered by the Farmers Home Administration. The basic justification for federal financing stems from rapid energy development as national policy warranting responsibility of the nation as a whole.

Alternatively, taxes can be levied on energy resource production within states for redistribution to areas with impacts. Severance taxes on coal in effect in Montana, North Dakota, and Wyoming effectively pass the cost of impact mitigation to consumers of the energy rather than to state taxpayers (Bronder, Carlisle, and Savage 1977). Severance tax collections from oil, gas, and coal for energy-exporting states are substantial (Table 17). In Louisiana, Texas, and Oklahoma, these taxes provided approximately one-fifth of the states' total 1979 tax collections.

Coal-producing states are likely to fare very well in this regard in the future. Kentucky increased its severance tax collections by $117 million to over $154 million from 1973 to 1978, almost exclusively on coal taxes. Coal states have recently increased their taxes, and some added taxes for energy conversion. Montana now has a

TABLE 17 STATE ENERGY SEVERANCE TAX COLLECTIONS, 1979

	REVENUES ($1000)			PERCENT TOTAL STATE REVENUES
	OIL/GAS	COAL	TOTAL	
Louisiana	500,666	—	500,666	22.3
Texas	1,021,017	—	1,021,017	17.8
Kentucky	404	153,613	154,017	7.4
Oklahoma	280,982	—	280,982	18.5
West Virginia	245	9,030	9,275	<1
New Mexico	138,511	—	138,511	16.4
Wyoming	308	30,278	30,586	8.9
Montana	8,208	42,049	50,257	12.5
Kansas	1,097	—	1,097	<1
North Dakota	13,533	11,970	25,503	7.9
Utah	6,175	—	6,175	<1

Source: U.S. Department of Commerce, Bureau of the Census 1980.

30 percent coal severance tax that was ruled legal by the Supreme Court despite challenges by electric utilities. Over 50 percent of this tax is allocated to the state's general fund, much of which will be used for education. Thirty-five percent of these revenues are to be used to mitigate coal development impacts.

Application fees for coal conversion facilities will also add revenues to states where these facilities are sited. North Dakota, for example, requires an application fee of $150,000 for a large energy facility; Montana requires about $2 million for a $1 billion energy facility. Thus, it appears that energy-consuming states will continue to pay higher prices for energy, and a flow of revenues should continue from energy-consuming to energy-producing states.

The taxation approach to mitigation implies that energy consumers should pay the full cost rather than be subsidized by taxpayers. However, this policy could penalize customers of energy firms on which they are dependent. Electric and gas utilities, in particular, have monopolies over defined market areas.

Finally, energy developers can provide funds, expertise, and assistance *in kind* to communities affected by a firm's activities (Richards 1978). Assistance from developers can also be made obligatory through state siting legislation (as in Wyoming) that provides for permit approvals conditional on impact mitigation measures (Valeu 1977). Such a siting procedure allows a state to determine whether an energy project fits into overall state development plans. This approach reflects a belief that energy firms should pay the full cost of energy development, but accepts that the extra costs will be passed on to consumers. The major differences between this policy and severance taxes are that the latter can provide funds for purposes other than direct impact mitigation and that entire states, rather than just impact areas, receive some financial benefit.

A combination of these three policy measures is likely to become common as energy resources are developed at a more intensive pace. State severance taxes allow states to get some longer-term benefit from non-renewable resources and to allocate funds to the areas which are most affected. At the same time, energy developers are under increasing pressure to pay for social as well as economic costs of energy development. Finally, the federal government may need to take a more responsible role in accounting for the cost of obtaining domestic substitutes for imported oil.

This chapter reviewed only a sample of possible approaches to help solve energy facility siting problems. Promising approaches are being developed to address site selection, location planning, and licensing aspects of the facility siting process. The question is whether these approaches are implemented successfully. In order to accomplish this, it is necessary for the parties involved to allow alternative procedures to be used. Utilities must be willing to open their planning to government agencies and interested parties and incorporate a range of goals into the site selection process. Also, there must be an attempt on the part of all parties to establish their siting priorities and communicate in good faith. The final chapter presents some of the implications of siting patterns and their possible implications for future energy supply.

7

Guiding Energy Supply Through the End of the Fossil-Fuel Era

Most coal advocates view the resource as an important interim fuel that can sustain continued economic growth as energy supply is diversified toward an emphasis on flow resources. Coal, with severe problems at every stage of its extraction and use, may not be the energy source that should power a twenty-first century economy. Environmental, safety, health, and logistical problems plague coal mining, transportation, and conversion. Local environmental impacts of coal use are overshadowed by global threats of increased carbon dioxide and acid rainfall. Since coal is viewed as an interim fuel, it is very likely that the next generation of coal conversion facilities will be the *last* generation of facilities that are constructed to convert coal into a more usable form. This makes siting decisions for these facilities important because they will determine the distribution and intensity of related impacts, influence patterns of energy supply, and create new interregional energy dependencies that will last into the next century. Adjusting siting patterns may be important in distributing the impacts of energy development. What will be the significance of new patterns of energy supply?

Adjusting Patterns of Development

The pattern of energy development can be an important factor influencing the size, nature, and distribution of impacts. It has been suggested that clustering energy facilities into "energy parks" could limit the overall area and population affected by energy development. Such development patterns may reduce political opposition by carefully selecting a site in which local attitudes favor this type of development. On the other hand, siting a larger number of smaller facilities throughout the country may be more equitable, reduce transmission losses, and permit better matches between energy needs and supply. However, this strategy would result in a larger population being exposed to facilities and could delay siting of facilities where political opposition occurs.

The rationale favoring smaller, more dispersed energy facilities is that their impacts will be considerably less on any one area and thus would more easily meet federal and state regulations; they would not encounter the public opposition faced by large facilities; and they would be brought on-stream more quickly. This is in contrast to the present trend of proposing large energy projects. Two public utility commissions,

New York and California, have requested utilities to consider more numerous but smaller energy facilities instead of large projects to maintain supply commitments.

The issue of larger versus smaller and more dispersed energy facilities was investigated by Los Alamos National Laboratory and was reported in the journal *Environment* (Ford 1980; Champion and Williams 1980; Lorber 1980; and Champion and Ford 1980). The study specifically investigated the advantages of siting four 500 MWe facilities (composed of two 250 MWe units each) and one 250 MWe facility as compared to siting one 3000 MWe facility (composed of four 750 MWe units) in southeastern Utah. This region was chosen because of the siting difficulties encountered by the big power projects in the region, such as Kaiparowits. Recall that a 500 MWe facility, although small compared to a 3000 MWe plant, is indeed a large project, and would have been among the largest in the nation in the 1960s.

An immediate advantage of small facilities is that 2250 MWe in smaller units is equivalent to 3000 MWe in the larger units. Many studies indicate that larger facilities are less reliable than smaller facilities and thus nine 250 MWe units provide as much reliable generating capacity to a utility system as four 750 MWe units (Ford 1980:28-29). When a large unit is down, a much greater share of a utility system is affected. This requires a utility to have more capacity or larger facilities on reserve in case of possible service outages. The undesirable financial impacts of having excessive reserve margins is a separate, but serious issue (Shenon 1980; Parisi 1980).

The attributes of the small and large power plants were evaluated in the Los Alamos study (Table 18). The larger plant is more thermally efficient than the smaller and has lower capital costs. Because the small facility is less efficient in converting coal to electricity, its air emissions will be about 15 percent greater than those of the larger facility. However, the smaller facility is more reliable and can be approved and constructed in 60-96 months compared to the 108-156 months required for the 3000 MWe facility. In addition, the lower levels of residuals generated by a small facility can be

TABLE 18 TYPICAL ATTRIBUTES OF SMALL AND LARGE POWER PLANTS

ATTRIBUTE	SMALL PLANT	LARGE PLANT
Total Generating Capacity	500 MWe	3000 MWe
Generating Unit Size (Boiler-Turbine)	250 MWe	750 MWe
Siting-Permit Period	24-26 months	36-48 months
Construction Time	36-60 months	72-108 months
Construction Workforce	650 (peak)	2680 (peak)
Operating Workforce	85	500
Land Area	300-800 acres	1500-2500 acres
Heat Rate[a]	11,500 Btu/kWh	10,000 Btu/kWh
Forced Outage Rate[b]	10-12%	18-20%
Capacity Factor[c]	68%	57%
Capital Cost (1976$)	$444/kW	$385/kW

[a]Heat (from burning coal) required to generate one kilowatt hour of electricity.
[b]Percentage of time generating units are forced out of service
[c]Kilowatt hours of electricity actually generated as a percentage of electricity that the unit would generate if operated continuously at 100% capacity.

Source: Ford 1980:26. Reproduced by permission. This article originally appeared in *Environment,* March 1980 (Vol. 22, No. 2), Heldref Publications, 4000 Abermarle St., N.W., Washington, D.C. 20016.

more easily accommodated by the environment than those produced by a large facility. As summarized by Champion and Williams (1980:31):

> Utilities might be wise to consider smaller power plants as a means of reducing environmental impact and thereby blunting opposition to new facilities. Small plants also offer a more substantive advantage when it comes to locating and gaining approval of power plant sites.

Instead of dispersing energy development throughout the country, energy facilities can be concentrated into a few locations. Such sites are commonly known as "power parks," "fossil energy centers," or "energy parks" and could be designed to generate as much as 10,000 MWe of electricity plus synthetic fuels. The aim of such proposals is to concentrate impacts into a localized area which would affect only a small number of people. The local population would bear a disproportionate share of the impacts, but the overall population would be spared most of the undesirable impacts of these facilities. Such a concentration of energy facilities has distinct benefits and disadvantages.

Proponents of energy parks maintain that the licensing of a single large site would be easier than securing separate licenses for dispersed facilities (Cirillo et al. 1976:348). Although the licensing effort will be time consuming and complicated, it is believed to be less than the licensing effort required for siting separate facilities at many dispersed sites. Likewise, it is anticipated that energy parks can use standard engineering and construction methods to reduce construction and delivery time schedules (Federal Energy Administration 1975:3). They will enjoy some localization economies. For instance, specialized service firms will be attracted to the area to provide prompt and efficient service and maintenance. It is also believed that energy parks will have a more stabilized construction force, reducing "boom and bust" problems for local communities and increasing productivity. The available waste heat could be an attractive source of inexpensive process steam for industry which would add more revenues and jobs to the area, expanding the economy (Federal Energy Administration 1976:28).

These energy centers are also likely to encounter some very serious problems. A study of the possibilities of developing energy parks in Pennsylvania identified some of the significant problems likey to be encountered (Ferrar et al. 1975). The study notes that energy parks may not reduce the cost of electricity; the concentration of energy facilities may actually increase land commitments to power generation; and local environmental problems are likely to be quite intense. A study prepared by Battelle Pacific Northwest on "energy centers" noted these and other problems (Federal Energy Administration 1975:3-5). Local socioeconomic effects could be quite serious; local resources (such as water) might be overtaxed by such developments; and it may actually be more difficult to license energy centers than it would be to license dispersed facilities. Many Air Quality Control Regions could not accept the increment of air pollution degradation accompanying energy centers, although a smaller single facility could be accommodated. Even in the East few sources are available that could provide the volumes of cooling water needed for a center projected to generate in excess of 10,000 MWe. The Pennsylvania study found that most sites for energy parks would require the construction of reservoirs (Ferrar et al. 1975:15):

> Since approximately 250 to 300 cubic feet of water per second will be evaporated by an energy park of 10,000 megawatts, no Pennsylvania river could meet this demand without a reservoir, even if ten percent of the ten-year low flow is the restricted consumption rate.

Despite these shortcomings, political realities may result in the development of energy parks in the future. Industry spokesmen have already announced that some regions of the country should be designated "energy zones" where environmental regulations should be relaxed. This would indeed be necessary for energy parks incorporating coal conversion facilities.

Regional Shifts in Energy Supply

Regional shifts in electricity supply resulting from facility siting decisions could occur because projected capacity increases are large; facilities are larger than in previous generations; and projected oil burning and nuclear facilities will probably not come on-line as planned. Thus, the outcome of a few siting decisions could be important in determining which regions are the princial energy suppliers. Figure 5 showed the location of announced coal conversion facilities, indicating that middle western and western states may become important sources of electricity supply. Siting patterns could be instrumental in increasing the economic fortunes of states supplying coal-generated energy at the expense of those states which must rely on external sources. Recent studies have shown that there has been a shift in real income from energy-consuming to energy-producing states (Miernyk 1978). Part of this shift will be reflected in increased tax revenues in energy producing states. However, the economic impacts of increased coal utilization will extend much further than the revenues to be collected through severance taxes. The National-Regional Impact Evaluation System (NRIES), a model developed by the U.S. Bureau of Economic Analysis, has measured the regional economic and demographic effects of advanced coal production in the U.S. In evaluating the regional growth impacts of a projected 1985 national coal production of 1,050 million tons, Wendling and Ballad (1980) found that statewide coal-related growth depends on the manufacturing base of the state and its coal resources or proximity to coal-producing regions. They note that Illinois will register the highest growth impacts with a swing toward coal:

> This is not surprising since the state benefits from all stages of coal development. First, it is one of the states where high-level, advanced coal development is assumed. It has a very high manufacturing durables base and is a source not only for mining equipment, but also for construction-related equipment. Finally, Illinois has a highly developed transportation industry which will be available to move the additional coal from the major areas of advanced coal development. . . . (Wendling and Ballard 1980:14)

However, the authors emphasize (1980:16) that on a per capita basis, the cumulative personal income increases from increased coal utilization to a 1985 level of 1,050 million tons will be $1,211 Wyoming, $714 in West Virginia, $274 in Montana, $262 in Kentucky, and $136 in Illinois. States such as Wyoming, Montana, and West Virginia will enjoy such high levels of per capita income increases since they are so sparsely populated.

Regional growth effects of increased coal utilization will depend on location decisions for coal facilities — a fruitful area of study for geographers. Information on where new coal facilities are to be sited, in concert with knowledge of their capital, labor, and resource inputs, can provide a better understanding of the regional growth impacts of a coal future.

Conclusion

In this book, we have broadly addressed the energy siting problem, focusing on some critical locational concerns. We have emphasized that the energy facility siting problem is related to many energy development issues. Adjusting energy facility locations, as well as their scale, influences the nature and intensity of impacts and benefits. In addition, political considerations make siting more difficult in some areas than in others. As a broad topic, siting encompasses the expertise of many disciplines. The perspective of the geographer, who is able to evaluate the interrelationships of various factors impinging upon a location, is essential in planning the location of energy facilities. As a generalist able to deal with problems of location and scale, the geographer provides a synthesis derived from the confusion of contradictory details in addressing a difficult siting problem. The specialist tools of geography have proven valuable in determining specific facility locations and in assessing the implications of energy facility patterns.

Bibliography

Atomic Industrial Forum. 1974. *Nuclear Power Plant Siting — A Generalized Process.* New York: Atomic Industrial Forum.
Averitt, P. 1973. "Coal," pp. 133-142 in D.A. Brobst and W.P. Pratt (editors), *United States Mineral Resources,* U.S. Geological Survey Professional Paper 820. Washington, DC: U.S. Government Printing Office.
Baram, M.S. 1976. *Environmental Law and Siting of Facilities: Issues in Land Use and Coastal Zone Management.* Cambridge, MA: Ballinger Press.
Bartlett, A.A. 1978. "Forgotten Fundamentals of the Energy Crisis," *American Journal of Physics,* Vol. 46, pp. 876-888.
Beatty, H. 1973. "Federal Water Pollution Control in Transition," p. 497 in Rocky Mountain Mineral Law Foundation (editors), *Rocky Mountain Mineral Law Institute: Proceedings of the 18th Annual Institute.* New York: Matthew Bender.
Benson, D.C. and F.J. Doyle. 1978. Projects to Expand Fuel Sources in Eastern States — An Update of Information Circular 8725, U.S. Bureau of Mines *Information Circular* 8765. Washington, DC: U.S. Government Printing Office.
Blair, J.M. 1976. *The Control of Oil.* New York: Pantheon Books.
Blake, R.R. and J.S. Mouton. 1964. *The Managerial Grid.* Houston, TX: Gulf Publishing Company.
Bosselman, F. and D. Callies. 1972. *The Quiet Revolution in Land Use Control.* Washington, DC: U.S. Government Printing Office.
Bronder, L.D., N. Carlisle, and M.D. Savage. 1977. *Financial Strategies for the Alleviation of Socioeconomic Impacts in Seven Western States.* Denver, CO: Western Governors' Regional Energy Policy Office.
Brunner, R.D. 1980. "Decentralized Energy Policies," *Public Policy,* Vol. 28 (March), pp. 71-91.
Business Week. **1976.** "The Power-Plant War on a Utah Plateau," No. 2415, (January 19), p. 22.
Business Week. **1977.** "The New Opposition to High Voltage Lines," November 7, p. 27.
Caldwell, L.K., L.R. Hayes, and I.M. MacWhirter. 1976. *Citizens and the Environment: Case Studies in Popular Action.* Bloomington, IN: Indiana University Press.
California Energy Resources Conservation and Development Commission. 1977. *California Energy Trends and Choices,* Vol. 7, Power Plant Siting. Sacramento, CA.
Calvert, J.R. 1978. "Licensing Coal-Fired Power Plants," *Power Engineering,* Vol. 82 (January), pp. 34-42.
Calzonetti, F.J. 1979. *Impacts of the Resource Conservation and Recovery Act on the Siting of Coal Conversion Facilities in the United States.* Oak Ridge, TN: Oak Ridge National Laboratory.
Calzonetti, F.J., M.E. Eckert, and E.J. Malecki. 1980. "Siting Energy Facilities in the USA: Policies for the Western States." *Energy Policy,* Vol. 7 (June), pp. 138-152.
Calzonetti, F.J. and G.A. Elmes. 1981. "Metal Recovery from Power Plant Ash: An Ecological Approach to Coal Utilization," *Geojournal* (forthcoming).

Carasso, M., et al. 1975. *The Energy Supply Planning Model.* San Francisco, CA: Bechtel Corporation.
Carlson, R., D. Freedman, and R. Scott. 1979. "A Strategy for a Non-Nuclear Future," *Environment,* Vol. 20, pp. 6-37.
Carnduff, S.B. 1981. "Environmental Mediation in a Federal Agency," pp. 48-55 in P.A. Marcus and W.H. Emrich (editors), *Environmental Conflict Management.* New York: American Arbitration Association, Clark-McGlennon Associates, Inc.
Carnon, L. and J. Kotkin. 1979. "Old Frontier Sees Bright New Future: West, Seeking to Control Its Destiny, Fights Colonial Image," *Washington Post* (June 17), p. A-1, 20.
Carter, L.J. 1978. "The Attorney General and the Snail Darter," *Science,* Vol. 200 (May 12), p. 628.
Champion, D.B. and A. Ford. 1980. "A New Look at Small Power Plants: Boom-Town Effects," *Environment,* Vol. 22 (June), pp. 25-31.
Champion, D.B. and M.D. Williams. 1980. "A New Look at Small Power Plants: The Environmental Effects," *Environment,* Vol. 22 (April), pp. 25-32.
Chisholm, M. 1971. "In Search of a Basis for Location Theory," *Progress in Geography,* Vol. 3, pp. 113-133. London: Edward Arnold.
Cirillo, R.R., et al. 1976. *An Evaluation of Regional Trends in Power Plant Siting and Energy Transport.* Argonne, IL: Argonne National Laboratory.
Cirillo, R.R., et al. 1977. *An Evaluation of Regional Trends in Power Plant Siting and Energy Transport.* Argonne, IL: Argonne National Laboratory.
Clark, P.B. and F.H. Cummings, Jr. 1981. "Selecting An Environmental Conflict Management Strategy," pp. 10-33 in P.A. Marcus and W.M. Emrich, (editors), *Environmental Conflict Management.* New York: American Arbitration Association, Clark-McGlennon Associates, Inc.
Coal Week. 1980. "DOE Synfuel Plant Siting Studies Inadequate," Vol. 6, p. 1.
Cohen, A.S. and K.W. Costello. 1975. *Regional Energy Modeling: An Evaluation of Alternative Approaches.* Argonne, IL: Argonne National Laboratory.
Committee on Science and Technology, Subcommittee on Science, Research, and Technology. 1979. *Materials Policy and Solid Waste Management.* Washington, DC: U.S. Government Printing Office.
Commoner, B. 1977. *The Poverty of Power.* New York: Bantam.
Commoner, B. 1979. *The Politics of Energy.* New York: Alfred A. Knopf.
Congressional Research Service. 1977. *Project Interdependence: U.S. and World Energy Outlook Through 1990.* Washington, DC: U.S. Government Printing Office.
Daily Oklahoman. 1980a. "Government Opens Lands for Leasing," November 24, p. 24, (Oklahoma City).
Daily Oklahoman. 1980b. "High Court Overturns Fugitive Return Ruling," December 9, p. 6, (Oklahoma City).
Darmstadter, J. 1972. "Energy Consumption: Trends and Patterns," pp. 155-223 in S. Schuer (editor), *Energy, Economic Growth, and the Environment.* Baltimore: Johns Hopkins University Press.
Davis, T.P. 1976. "Citizen's Guide to Intervention in Nuclear Power Siting: A Blueprint for Alice in Nuclear Wonderland," *Environmental Law,* Vol. 6 (Spring), pp. 621-674.
Di Nunno, J.J. 1975. *Regional Approaches to Power Plant Siting in the United States of America.* Vienna: International Atomic Energy Agency, SM-188/42.
Dobson, J.E. 1979. "A Regional Screening Procedure for Land Use Suitability Analysis," *The Geographical Review,* Vol. 69 (April), pp. 224-234.
Eckert, M.E. 1977. "A Survey and Evaluation of State Energy Facility Siting Laws," (unpublished paper), Norman, OK: Science and Public Policy Program.
Energy Policy Staff, Office of Science and Technology, 1968. *Considerations Affecting Steam Power Plant Site Selection.* Washington, DC: Executive Office of the President.

Energy Research and Development Administration. 1977. *Managing the Socioeconomic Impacts of Energy Development.* Washington, DC: U.S. Government Printing Office.
Faber, J.H. 1976. "U.S. Overview of Ash Production and Utilization," in *Ash Utilization.* Morgantown, WV: Morgantown Energy Research Center, pp. 5-13.
Fairfax, S.K. 1978. "A Disaster in the Environmental Movement," *Science,* Vol. 197 (February), pp. 743-748.
Federal Energy Administration. 1975. *Siting Energy Facilities at Camp Gruber, Oklahoma.* Washington, DC: U.S. Government Printing Office.
Federal Energy Administration. 1976. *Energy Facility Siting Workshops.* Washington, DC: Federal Energy Administration.
Farney, D. 1980. "Synfuels Bill of $20 Billion Passed by House," *Wall Street Journal* (June 27), p. 1.
Ferrar, T.A., et al. 1975. *Energy Parks and the Commonwealth of Pennsylvania — Issues and Recommendations.* University Park, PA: Center for the Study of Environmental Policy.
Ford, A. 1980. "A New Look at Small Power Plants: Is Smaller Better?", *Environment,* Vol. 22, pp. 25-33.
Ford Foundation. 1974. *A Time to Choose.* Cambridge, MA: Ballinger Press.
Fradkin, P.L. 1977. "Craig, Colorado: Population Unknown, Elevation 6,185," *Audubon,* Vol. 79 (July), pp. 118-127.
Galbraith, J.K. 1967. *The New Industrial State.* Boston: Houghton-Mifflin Company.
Georgescu-Roegen, N. 1976. "Energy and Economic Myths," pp. 3-36 in *Energy and Economic Myths: Institutional and Analytical Economic Essays.* New York: Pergamon Press, Inc.
Gilmore, J.S. and M.K. Duff. 1975. *Boom Town Growth Management.* Boulder, CO: Westview Press.
Greenberg, M.R., R. Anderson, and G.W. Page. 1978. *Environmental Impact Statements.* Washington, DC: Association of American Geographers, *Resource Paper 78/3.*
Greenhut, M.L. 1951-52. "Integrating the Leading Theories of Plant Location," *Southern Economic Journal,* Vol. 18, pp. 526-538.
Greenhut, M.L. 1963. *Microeconomics and the Space Economy.* Chicago: Scott Foresman.
Hillsman, E.L. and D.R. Alvic. 1980. "Patterns of 'Outsiting' by U.S. Electric Utilities," (unpublished manuscript), Oak Ridge, TN: Oak Ridge National Laboratory.
Hobbs, B.F. 1980. "Multiobjective Power Plant Siting Methods," *Journal of the Energy Division,* Vol. 106 (October), pp. 187-200.
Holden, C. 1977. "Contract Archeology: New Source of Support Brings New Problems," *Science,* Vol. 196, pp. 1070-72.
Hopkins, L.D. 1977. "Methods for Generating Land Suitability Maps: A Comparative Evaluation," *Journal of The American Institute of Planners,* Vol. 43, pp. 386-400.
Hornblower, M. 1979. "The Sagebrush Revolution: Westerners Fight U.S. Restrictions on Federal Land," *The Washington Post* (June), p. B1, 3.
Hotelling, H. 1929. "Stability in Competition," *Economic Journal,* Vol. 34, pp. 41-57.
Hubbert, M.K. 1977. "World Oil and Natural Gas Reserves and Resources," pp. 632-644 in Congressional Research Service (editors), *Project Interdependence: U.S. and World Energy Outlook Through 1990.* Washington, DC: U.S. Government Printing Office.
Josephy, A.M. 1976. "Kaiparowits: The Ultimate Obscenity," *Audubon,* Vol. 78 (March), pp. 64-67.
Kash, D.E. et al. 1976. *Our Energy Future.* Norman, OK: University of Oklahoma Press.
Keeney, R.L. 1980. *Siting Energy Facilities.* New York: Academic Press.

Kirschten, J.D. 1977. "The Clean Air Conference — Something for Everybody," *National Journal,* Vol. 9 (August 13), pp. 1261-63.

Labys, W.C., S. Paik, and A.M. Liebenthal. 1979. "An Econometric Simulation Model of the U.S. Market for Steam Coal," *Energy Economics,* Vol. 1 (January), pp. 19-26.

Lamm, R.D. 1976. "States Rights vs. National Energy Needs," *Natural Resources Lawyer,* Vol. 9, pp. 41-48.

Lecomber, R. 1979. *The Economics of Natural Resources.* New York: John Wiley and Sons.

Lloyd, P.E. and P. Dicken. 1977. *Location in Space: A Theoretical Approach to Economic Geography.* Second Edition. New York: Harper and Row.

Lorber, H.W. 1980. "A New Look at Small Power Plants: Health and Safety," *Environment,* Vol. 22 (May), pp. 25-31.

Love, S. 1977. "The New Look of the Future," *The Futurist,* Vol. 11 (April), pp. 78-80.

Loving, R. Jr. 1980. "The Railroads' Bad Trip to Deregulation," *Fortune,* Vol. 102 (August 25), pp. 45-50.

Lovins, A.B. 1976. "Energy Strategy: The Road Not Taken," *Foreign Affairs,* Vol. 55, pp. 65-96.

Lovins, A.B. 1977. *Soft Energy Paths: Toward a Durable Peace.* Cambridge, MA: Friends of the Earth International.

Maher, E. 1977. "The Dynamics of Growth in the Electric Power Industry," pp. 149-216 in K. Sayre (editor), *Values in the Electric Power Industry.* Notre Dame, IN: University of Notre Dame Press.

Maier, P.M. and B.F. Hobbs. 1978. "The Locational Response to Regulatory Policy: A Regional Analysis of Energy Facility Location." Paper presented to the Annual Meeting of the Northeast Regional Science Association.

Marcus, P.A. 1981. "Designing Effective Scoping Meetings: Suggestions for Federal Compliance with the NEPA Regulations of 1978," pp. 56-66 in P.A. Marcus and W.H. Emrich (editors), *Environmental Conflict Management* (working paper series). New York: American Arbitration Association, Clark-McGlennon Associates, Inc.

Marshall, E. 1980. "Energy Forecasts: Sinking to New Lows," *Science,* Vol. 208 (June 20), pp. 1353-56.

McHarg, I. L. 1969. *Design with Nature.* New York: Natural History Press.

Messing, M., H.P. Friesema, and D. Morell. 1979. *Centralized Power.* Cambridge, MA: Oelgeschlager, Gunn and Hain Publishers, Inc.

Metz, William D. 1977. "Nuclear Licensing: Promised Reform Miffs All Sides of Nuclear Debate," *Science,* Vol. 198 (November 11), p. 590.

Metzer, J.E. and E.V. Stenehjem. 1977. "An Analysis of the Socio-economic Impacts of Alternative Siting Patterns: Mine-Mouth Versus Remote Electrical Generation." Paper presented at the 79th Annual Meeting of the Air Pollution Control Association, Toronto, Ontario.

Miernyk, W.H. 1978. "Regional Consequences of High Energy Prices in the United States," pp. 5-32 in W.H. Miernyk, F. Giarratani, and C.F. Socher (editors) *Regional Impacts of Rising Energy Prices.* Cambridge, MA: Ballinger Publishing Company.

Miller, M.W. and G.E. Kaufman. 1978. "High Voltage Overhead," *Environment,* Vol. 20 (January/February), pp. 6-15.

Mitchell, E.J. and P.R. Chatletz. 1975. *Toward Economy in Electric Power.* Washington, DC: American Enterprise Institute.

Montana Major Facility Siting Act of 1975. *Montana Revised Codes Annotated* § 70-802. Cumulative Supplement 1975.

Morgan, R. and S. Jerabek, 1974. *How to Challenge Your Local Electric Utility.* Washington, DC: Environmental Action Foundation.

Bibliography 67

Murdock, C. 1977. "Legal and Economic Aspects of the Electric Utility's 'Mandate to Serve,'" pp. 100-115 in K. Sayre (editor), *Values in the Electric Power Industry.* Notre Dame, IN: University of Notre Dame Press.
Myhra, D. 1977. "Fossil Projects Need Siting Help Too," *Public Utilities Fortnightly,* Vol. 100 (September), pp. 24-28.
National Electric Reliability Council. 1980. *Tenth Annual Review of Overall Reliability and Adequacy of the North American Bulk Power Systems.* Princeton, NJ: National Electric Reliability Council.
National Research Council. 1979. *Energy in Transition, 1985-2010.* San Francisco: W.H. Freeman and Company.
Natural Resources Defense Council. 1977. *Land Use Controls in the United States.* Washington, DC: Natural Resources Defense Council.
New England River Basins Commission and United States Geological Survey Resource Plannng Analysis Office. 1980. *Power Plant Siting: Overview Report.* (Unpublished manuscript).
***New Mexico Energy Resources Act.* 1975.** Chapter 289 of the New Mexico Laws of 1975. New Mexico Statutes Annotated. Section 8 B(3).
Norman, C. 1981. "Study Says Coal Cheaper than Nuclear," *Science,* Vol. 212, p. 652.
Novick, R. 1977. "Energy Round-Up," *Environment,* Vol. 19 (June-July), p. 37.
Office of Technology Assessment. 1978. *A Technology Assessment of Coal Slurry Pipelines.* Washington, DC: Office of Technology Assessment.
Organization for Economic Co-operation and Development. 1977. *Siting of Major Energy Facilities.* Paris: OECD.
Organization for Economic Co-operation and Development. 1980. *Siting Procedures for Major Energy Facilities.* Paris: OECD.
O'Riordan, T. 1976. *Environmentalism.* London: Pion Press.
Osleeb, J.P. and I.M. Sheskin. 1977. "Natural Gas: A Geographical Perspective," *The Geographical Review,* Vol. 67 pp. 71-85.
O'Toole, J. 1976. *Energy and Social Change.* Cambridge, MA: The MIT Press.
Parisi, A.J. 1980. "Utilities Have Cause to Thank Their Critics," *New York Times,* September 7.
Paslay, R. 1978. "$500 Million Plant to Locate Here," Morgantown *Morning Reporter,* July 7.
Pasztor, A. 1981. "Synfuels Plant Blasted for Cost Overruns and Poor Management in Federal Reports," *Wall Street Journal* (February 17).
Plummer, J.L. 1977. "The Federal Role in Rocky Mountain Energy Development," *Natural Resources Journal,* Vol. 17 (April), pp. 247-255.
President's Commission on Coal, 1980. *The Acceptable Replacement of Imported Oil with Coal.* Washington, DC: U.S. Government Printing Office.
Priest, J. 1969. *Principles of Public Utility Regulation.* Charlottesville, VA: Michie Press.
Quarles, J. 1978. "Impact of the 1977 Clean Water Act Amendments on Industrial Discharges," *Environment Reporter,* Vol. 8 (March 17), pp. 1-11.
Radian Corporation. 1976. *The Coal Resource System: Final Draft.* Austin, TX: Radian Corporation.
Raffle, B.I. 1978. "The New Clean Air Act — Getting Clean and Staying Clean," *Environment Reporter,* Vol. 8 (May 19), pp. 1-28.
Rattner, S. 1978. "Tough Task in Implementing Crippled Energy Plan," *New York Times National Economic Survey.* January 8, Secton 12, p. 13.
Reynolds, J.D. 1980. "Power Plant Cooling Systems: Policy Alternatives," *Science,* Vol. 207 (January 25), pp. 367-372.
Rich, C.H. 1978. *Projects to Expand Energy Sources in the Western States – An Update of Information Circular 8719,* U.S. Bureau of Mines *Information Circular* 8772. Washington, DC: U.S. Department of Interior.

Richards, W.R. 1978. "Mining Industry Housing and Community Relations." Paper presented to the American Mining Congress Coal Convention, St. Louis.
Richardson, H.E. 1969. *Regional Economics.* New York: Praeger Publishers.
Roberts, M.J. and J.S. Bluhm. 1981. *The Choices of Power: Utilities Face the Environmental Challenge.* Cambridge, MA: Harvard University Press.
Rock, J.M. 1977. "No Boomtown on the Kaiparowits Plateau: Who Made the Decision and Why?" *Intellect,* Vol. 105 (February), pp. 248-250.
Rowe, M.D. et al. 1979. *An Assessment of Nuclear Power Plant Siting Methods.* Upton, NY: Brookhaven National Laboratory.
Schanz, J.J., Jr., J.W. Sawyer, Jr., and H. Perry. 1979. *The Future Regional Distribution of National Synthetic Fuel Capacity.* Washington, DC: Resources for the Future.
Schumacher, E.F. 1973. *Small is Beautiful: Economics as if People Mattered.* London: Bland and Briggs.
Science and Public Policy Program. 1975. *Energy Alternatives: A Comparative Analysis.* Norman, OK: University of Oklahoma Press.
Scott, Allen J. 1971. *An Introduction to Spatial Allocation Analysis.* Washington, DC: Association of American Geographers, *Resource Paper* 9.
Scott, W.G. and D.K. Hart. 1979. *Organizational America.* Boston, MA: Houghton-Mifflin Company.
Sears, F.W., M.W. Zemansky, H.D. Young. 1980. *College Physics,* Fifth Edition. Reading, MA: Addison-Wesley Publishing Company.
Seley, J. and J. Wolpert. 1974. "A Strategy of Ambiguity in Locational Conflicts," pp. 275-300 in K.R. Cox, D.R. Reynolds, and S. Rokkan (editors), *Locational Approaches to Power and Conflict.* New York: John Wiley and Sons.
Shenon, Philip W. 1980. "Many Electric Utilities Suffer as Conservation Holds Down Demand," *Wall Street Journal* (October 9).
Shields, Scott. 1980. "Company Vows to Construct Coke Plant," Morgantown *Dominion Post* October 12.
Smith, D.M. 1971. *Industrial Location: An Economic Geographical Analysis.* New York: John Wiley and Sons, Inc.
Solow, R.M. 1973. "Is the End of the World at Hand?", pp. 39-61 in A. Weintraub, E. Schwartz, and J.R. Aronson (editors), *The Economic Growth Controversy.* White Plains, NY: International Arts and Sciences Press.
Southern Interstate Nuclear Board. 1976. *Power Plant Siting in the United States.* Atlanta, GA: Southern Interstate Nuclear Board.
Stobaugh, R. and D. Yergin. 1979. *Energy Future.* New York: Random House.
Stoker, H.S., S.L. Seager, and R.L. Capener, 1975. *Energy: From Source to Use.* Glenview, IL: Scott, Foresman, and Company.
Susskind, L.E. and S.R. Cassella. 1980. "The Dangers of Preemptive Legislation: The Case of LNG Facility Siting in California," *Environmental Impact Assessment Review,* Vol. 1 (March), pp. 9-26.
Thomas, K.W. 1976. "Conflict and Conflict Management," pp. 889-935 in M.D. Dunnette (editor), *Handbook of Industrial and Organizational Psychology.* Chicago: Rand McNally.
Thomas, M. 1980. "Explanatory Frameworks for Growth and Change in Multiregional Firms," *Economic Geography,* Vol. 56, pp. 1-17.
Thwaites, A.T. 1978. "Technological Change, Mobile Plants, and Regional Development," *Regional Studies,* Vol. 12, pp. 445-461.
U.S. Congress, Office of Technology Assessment. 1978. *A Technology Assessment of Coal Slurry Pipelines.* Washington, DC: Office of Technology Assessment.
U.S. Congress, Senate Committee on Energy and Natural Resources. 1977. *National Energy Transportation, Vol. 1: Current Systems and Movements.* Committee Print by the Congressional Research Service, Washington, DC: U.S. Government Printing Office.

U.S. Department of Commerce, Bureau of the Census. 1980. *State Tax Collections in 1979.* Washington, DC: U.S. Government Printing Office.
U.S. Department of Energy, 1980. *A Draft Environmental Impact Statement for a Solvent-Refined Coal Plant in Morgantown, West Virginia.* Washington, DC: U.S. Government Printing Office.
U.S. Department of Energy, 1981. *Solvent Refined Coal – II, Demonstration Project.* Final Environmental Impact Statement. Washington, DC: U.S. Department of Energy.
U.S. Department of Energy, Economic Regulatory Administration, Division of Power Supply and Reliability. 1980. *Electric Power Supply and Demand for the Contiguous United States 1980-1989.* Washington, DC: U.S. Department of Energy.
U.S. Department of Energy, Energy Information Administration. 1979. *Inventory of Power Plants in the United States.* Washington, DC: U.S. Government Printing Office.
U.S. Department of Energy, Energy Information Administration. 1980a. *Statistics of Privately Owned Electric Utilities in the United States – 1979.* Washington, DC: U.S. Department of Energy.
U.S. Department of Energy, Energy Information Administration. 1980b. *Energy Data Report.* Washington, DC: Energy Information Office.
U.S. Department of Energy, Energy Information Administration. 1981. *Monthly Energy Review.* Washington, DC: U.S. Government Printing Office.
U.S. Department of the Interior, Bureau of Reclamation, Upper Missouri Region. 1977. *ANG Coal Gasification Company, North Dakota Project: Draft Environmental Impact Statement.* Billings, MT: Bureau of Reclamation.
U.S. Regulatory Council. 1980. "Proposed Regulations on Surface Coal Mining and Processing Plants," *Federal Register,* Vol. 45. No. 228 (November 24), Part II, p. 77759.
Valeu, R.L. 1977. "Financial and Fiscal Aspects of Monitoring and Mitigation." Paper presented to the Symposium on State-of-the-Art Survey of Socioeconomic Impacts Associated with Construction/Operation of Energy Facilities, St. Louis, Missouri, January.
Ward, J.E. 1979. "Nuclear Power's Potential," pp. 57-65 in American Society of Civil Engineers, National Energy Policy Committee (editors), *Electric Power Today: Problems and Potential.* New York: American Society of Civil Engineers.
Warren, R. 1978. Address to the Conference on the Urban Coast and Energy Alternatives, Princeton, NJ, May 1978.
Weber, A. 1929. *Theory of the Location of Industries.* (Translated by C.J. Friedrich). Chicago: University of Chicago Press.
Weeter, D.M. and H. Carson. 1979. *Technical Aspects of the Resource Conservation and Recovery Act.* Oak Ridge, TN: Oak Ridge National Laboratory.
Wendling, R.M. and K.P. Ballard. 1980. "Projecting the Regional Economic Impacts of Energy Development," *Growth and Change,* Vol. 11, pp. 7-17.
White, I.L. et al. 1977. *Energy from the West: A Progress Report of a Technology Assessment of Western Energy Resource Development, Vol. 1, Summary.* Washington, DC: U.S. Environmental Protection Agency.
White, I.L. et al. 1978. *Energy From the West: "Draft" Policy Analysis Report.* Washington, DC: U.S. Environmental Protection Agency.
White, I.L. et al. 1979. *Energy From the West: Policy Analysis Report.* Washington, DC: U.S. Environmental Protection Agency.
Wilson, C.L. 1980. *Coal — Bridge to the Future.* Cambridge, MA: Ballinger Publishing Company.
Winter, J.V. and D.A. Conner. 1978. *Power Plant Siting.* New York: Van Nostrand Reinhold Company.

Wolpert, J. 1976. "Regressive Siting of Public Facilities," *Natural Resources Journal*, Vol. 16 (January), pp. 103-115.
Young, L.B. 1973. *Power Over People.* New York: Oxford University Press.